Lecture Notes in Economics and Mathematical Systems

676

More information about this series at
http://www.springer.com/series/300

Frédéric Amblard • Francisco J. Miguel •
Adrien Blanchet • Benoit Gaudou
Editors

Advances in Artificial Economics

Springer

Editors
Frédéric Amblard
Adrien Blanchet
Benoit Gaudou
University Toulouse 1 Capitole
Toulouse
France

Francisco J. Miguel
Laboratory for Socio-Historical Dynamics (LSDS)
Universitat Autònoma de Barcelona
Cerdanyola del Vallès
Spain

ISSN 0075-8442 ISSN 2196-9957 (electronic)
ISBN 978-3-319-09577-6 ISBN 978-3-319-09578-3 (eBook)
DOI 10.1007/978-3-319-09578-3
Springer Cham Heidelberg New York Dordrecht London

Library of Congress Control Number: 2014955224

Printed on acid-free paper

Springer is part of Springer Science+Business Media (www.springer.com)

Preface

The interactions between Computer Science and the Social Sciences have grown fruitfully along the past 20 years. The mutual benefits of such a cross-fertilization stand as well at a conceptual, technological or methodological level. Economics in particular benefited from innovations in multi-agent systems in Computer Science leading to agent-based computational economics and in return the multi-agent systems benefited for instance of economic researches related to mechanisms of incentives and regulation to design self-organized systems.

Created 10 years ago, in 2005 in Lille (France) by Philippe Matthieu and his team, the Artificial Economics conference series reveals the liveliness of the collaborations and exchanges among computer scientists and economists in particular. The excellent quality of this conference has been recognized since its inception and its proceedings have been regularly published in Springer's Lecture Notes in Economics and Mathematical Systems series. At about the same period, the European Social Simulation Association was created and decided to support an annual conference dedicated to computational approaches of the social sciences. Both communities kept going alongside for the past ten years presenting evident overlaps concerning either their approaches or their members. This year, both conferences have decided to join their efforts and hold a common conference, Social Simulation Conference, in Barcelona, Spain, 1st to 5th September 2014 which will host the 10th edition of the Artificial Economics Conference.

In this edition, 32 submissions from 11 countries were received, from which we selected 20 for presentation (near 60 % acceptance). The papers have then been revised and extended and 19 papers were selected in order to make part of this volume.

We are very grateful to the authors of the submissions who provided the basic material of the conference, i.e. original and interesting research articles. We are also very grateful to all the members of the Program Committee and the additional reviewers for their hard work. Thanks are also due to Paolo Pellizari, Andrea Teglio,

Tim Verwaart and Friederike Wall (members of the AE Steering Committee) and Flaminio Squazzoni (ESSA President).

Toulouse, France — Frédéric Amblard
Cerdanyola del Vallès, Spain — Francesc J. Miguel
Toulouse, France — Benoit Gaudou
Toulouse, France — Adrien Blanchet
Summer 2014

Acknowledgements

We would like to thank all the members of the Scientific Committee for their participation in the selection of enclosed papers:

Frédéric Amblard (Chair)—IRIT, Université Toulouse 1 Capitole, France
Simone Alfarano—Universidad Jaume I de Castellón, Spain
Javier Arroyo—University Complutense de Madrid, Spain
Bruno Beaufils—LIFL, Université Lille 1, France
Doris Behrens—Alpen-Adria Universität, Austria
Olivier Brandouy—University of Paris 1, France
Charlotte Bruun—Aalborg University, Denmark
Andrea Consiglio—Università degli Studi di Palermo, Italy
Giorgio Fagiolo—Scuola Superiore Sant'Anna, Italy
José Manuel Galán—Universidad de Burgos, Spain
Lynne Hamill—University of Surrey, United Kingdom
Florian Hauser—Universität Innsbruck, Austria
Cesáreo Hernández—Universidad de Valladolid, Spain
Jean Daniel Kant—University of Paris 6, France
Stephan Leitner—Alpen-Adria-Universität, Klagenfurt, Austria
Marco Licalzi—Università "Ca' Foscari" di Venecia, Italy
Philippe Mathieu—LIFL, Université Lille 1, France
Mishael Milakovic—University of Bamberg, Germany
Akira Namatame—National Defense Academy, Japan
Sjoukje Osinga—Wageningen University, The Netherlands
Paolo Pellizzari—Università "Ca' Foscari" di Venecia, Italy
Marta Posada—Universidad de Valladolid, Spain
Marco Raberto—Università di Genova, Italy
Juliette Rouchier—CNRS, France
Enrico Scalas—Università del Piemonte Orientale, Italy
Klaus Schredelseker—Universität Innsbruck, Austria
Flaminio Squazzoni—University of Brescia, Italy
Andrea Teglio—Universidad Jaume I de Castellón, Spain

Elpida Tzafestas—University of Athens, Greece
Tim Verwaart—Wageningen University, The Netherlands
Friederike Wall—Alpen-Adria-Universität Klagenfurt, Austria
Murat Yildizoglu—Univ. Montesquieu Bordeaux IV, France

Contents

Does Collaboration Pay? An Investigation for the Domain of Distributed Investment Decisions

Stephan Leitner, Alexander Brauneis, and Alexandra Rausch

1 Introduction

One of the most important tasks of corporate financial management is to assure the efficiency of investment decisions (Ryan and Ryan 2002; Dutta and Fan 2009). In corporate practice it can be observed that departments are often endowed with decision making authority regarding their investments. This is due to rapidly changing markets, products, and technologies, and decentral managers who are usually better informed with respect to these volatile economic circumstances (Schuster and Clarke 2010). Thus, on the one hand, decentralizing decision-making brings along the opportunity to evaluate investment opportunities on a more sound basis. On the other hand, it evokes the need of coordination of the distributed investment decisions. What adds complexity on top is that the departments' interests are often divergent which is why uncoordinated distributed decision making usually results in not achieving the corporate objective (e.g. shareholder value maximization) (Young and O'Byrne 2002; Baldenius 2003; Arya et al. 1996; Kouvelis and Lariviere 2000). Furthermore, corporate investment budgets are usually limited and investment opportunities often compete for this limited amount of financial resources (Baldenius et al. 2007). Hence, it will only be possible to fund a subset of profitable investment opportunities. In order to assure the efficiency of distributed investment decision making (in terms of value maximization), well developed "decision-making rules and procedures" turn out to be essential. Such rules and procedures are usually embodied in coordination mechanisms which align the decentral decisions to the overall corporate objective (Georgiades et al. 2000; Baldenius et al. 2007; Leitner and Behrens 2013, 2014).

S. Leitner (✉) • A. Brauneis • A. Rausch
Faculty for Business and Economics, Alpen-Adria Universität Klagenfurt, Klagenfurt, Austria
e-mail: stephan.leitner@aau.at; alexander.brauneis@aau.at; alexandra.rausch@aau.at

F. Amblard et al. (eds.), *Advances in Artificial Economics*, Lecture Notes in Economics and Mathematical Systems 676, DOI 10.1007/978-3-319-09578-3_1

In our model, we utilize hurdle rates for coordination purposes. Hurdle rates are referred to as the required rate of return of an investment opportunity (Dutta and Fan 2009), i.e., hurdle rates are usually specified as departmental capital charges for opportunity costs. The concept of hurdle rates seems to be related to the concept of internal transfer pricing such as that hurdle rates can be regarded as the particular price which a department gets charged for the invested (or borrowed) amount of money. However, the standard textbook case suggests to fix hurdle rates equal to the cost of capital. This might be sufficient for the case of unlimited financial resources, but does not provide a solution for the case of multiple investment opportunities which compete for the same pot of funding. Baldenius et al. (2007) take account of this circumstance and introduce the concept of the Competitive Hurdle Rate (CHR) mechanism. The CHR mechanism is a capital budgeting mechanism for coordinating distributed investment decisions which considers investment opportunities that are mutually exclusive due to scarce financial resources.

Based on divisional reports about the characteristics of projects CHR derives cost charges which departments get charged in every subsequent time period in the case that they decide to operate an investment opportunity. These cost charges comprise depreciation and capital charges, whereby the capital charges are based on hurdle rates which are competitively (and similarly to a second price auction) computed on the basis of all divisional reports. Baldenius et al. (2007) derive the CHR mechanism from an agency model and show that—under specific premises—it is strongly incentive compatible. Such agency models usually assume fully rational, perfectly homogenous, and non interacting agents. Axtell (2007) refers to these core assumptions as the“neoclassical sweetspot”. In most cases, these assumptions are made in order to assure analytical tractability (Kirman 1993; Irlenbusch 2006; Leitner 2012). Hendry (2002) reasons another feature which is usually incorporated into agency models, i.e., the agents' full competence in carrying out tasks. However, given this large set of assumptions, it is plausible to assume that outcomes of agency models' are sensitive to changes in their basic frameworks. Leitner and Behrens (2013, 2014) take account of the very strict "neoclassical assumptions", and assume away full rationality, perfect homogeneity (regarding departments as well as regarding investment opportunities), and the department managers' perfect competence in forecasting, which are incorporated in the CHR model (Baldenius et al. 2007). By doing so, they investigate the robustness and applicability of the CHR mechanism to real world situations outside the "neoclassical sweetspot". However, Leitner and Behrens (2013, 2014) hold up the assumption that projects are carried out by exactly one organizational unit and simply assume away spillover effects. In this paper, we relax intraorganizational boundaries and allow for cooperation among departments. In particular, we adapt the CHR mechanism proposed by Baldenius et al. (2007) to situations in which projects that are carried out *jointly* by organizational departments compete for scarce financial resources. This is a novelty. Moreover, we investigate the robustness of our adapted mechanism to situations that are closer to real world situations. As suggested by Leitner and Behrens (2013, 2014), we consider limited rationality, heterogeneity with respect to investment opportunities and departments, and a certain level of incompetence in forecasting measures associated with investment opportunities. By comparison to the case of efficiently made investment decisions (in terms of value maximization), we narrowly characterize efficiency losses due to distributed investment decisions. Moreover, we conduct a comprehensive sensitivity analysis which allows for insights into the dynamics of coordinating decentralized and autonomous investment decisions. Since hurdle rate based coordination mechanisms as well as transfer price based coordination mechanisms seem to be frequently utilized in corporate practice (Baldenius et al. 2007), a further investigation of the robustness of such mechanisms appears highly relevant, also from the practitioners' point of view. Our work complements previous research on hurdle rates in the context of corporate investment decisions (Baldenius 2003; Baldenius et al. 2007) as well as work on the robustness of such coordination mechanisms to situations that are closer to real world situations (Leitner and Behrens 2013, 2014).

2 Simulation Model

We model organizations to consist of at least 4 departments i ($i = 4, \ldots, m$) and one coordinating unit. At $t = 0$ departments are in charge of proposing investment projects, J ($j = 2, \ldots, n$) to the coordinating unit. Each investment project is carried out by $z = m/n \geq 2$ departments, whereby each department autonomously decides whether or not to carry out the project. The function $f(i, j)$ represents whether ($f(i, j) = 1$) or not ($f(i, j) = 0$) department i is involved in project j. Financial resources are scarce which is why at most one project can be funded. The organization aims at maximizing its shareholder value (SHV), while departments aim at maximizing their individual utilities.

The following measures are associated with investment projects: (1) an initial cash outlay, κ_j, necessary to launch the investment project; (2) an intertemporal distribution of cash flows, $\mathbf{x}_{ij} = [x_{ij1}\ x_{ij2}\ \ldots\ x_{ijT}]$; (3) an efficiency parameter in operating the jointly proposed project per department, ρ_i,[1] scaling the cash flows such that the present value PV_{ij} is given by:

$$PV(\mathbf{x}_{ij}, \rho_i, r) := \mathbf{x}_{ij} \circ \mathbf{r}(r) \cdot \rho_i \ , \tag{1}$$

where $\mathbf{r}(r) = [(1 + r)^{-1}\ \ldots\ (1 + r)^{-T}]$ denotes the vector of discount factors. Consequently, project j's net present value (NPV) results in $\Lambda_j(r, \mathbf{p}_j) := \sum_{\forall i: f(i,j)=1} PV_{ij}(\mathbf{x}_{ij}, \rho_i, r) - \kappa_j$, where $\mathbf{p}_j = [\rho_{i1}\ \ldots\ \rho_{iz}]$, $\forall i : f(i, j) = 1$ represents the efficiency parameters of all departments which are involved in project j, and r denotes an interest rate.

[1]As each department is involved in exactly one investment project, we can suppress the notion of j.

For each investment opportunity the coordinating unit calculates hurdle rates $r_1^*, \ldots, r_j^*, \ldots, r_n^*$ according to the procedure introduced in Leitner and Behrens (2013, 2014). Whenever departments decide to put their proposed projects into action, they will be charged the hurdle rate for the initial cash outlay. To do so, for each investment project j we compute the highest NPV of all projects other than j, i.e., $\Lambda_j^* = max\{\Lambda_1(r_c, \mathbf{p}_1), \ldots, \Lambda_{j-1}(r_c, \mathbf{p}_{j-1}), \Lambda_{j+1}(r_c, \mathbf{p}_{j+1}), \ldots, \Lambda_n(r_c, \mathbf{p}_n)\}$, and a vector of reference efficiency parameters $\mathbf{p}_j^* := \Lambda_j^* \cdot \mathbf{p}_j / \Lambda_j(r_c, \mathbf{p}_j)$, where r_c denotes the corporation's cost of capital. The reference efficiency parameter, ρ_j^*, is the efficiency level at which, *ceteris paribus*, project j is at least as profitable as Λ_j^* (in terms of NPV). Project j's hurdle rate results in the internal rate of return at the reference efficiency level, i.e., $\Lambda_j(r^*j, \mathbf{p}_j) = 0$.[2]

As each department autonomously decides whether or not to carry out an investment project, we need to derive an *incentive compatible* mode of allocating the initial cash outlay, κ_j, to the departments. We compute department i's share on κ_j according to

$$\lambda_{ij} = \frac{PV(\mathbf{x}_{ij}, \rho_j^*, r_j^*)}{\sum_{\forall k: f(k,j)=1} PV(\mathbf{x}_{kj}, \rho_k^*, r_j^*)} \tag{2}$$

Next, the coordinating unit announces the hurdle rates and the shares of the initial cash outlay to the departments. On the basis of this information, each department decides whether ($I_{ij} = 1$) or not ($I_{ij} = 0$) to put the proposed project into action

$$I_{ij} = \begin{cases} 1, & \text{if} PV_{ij}\left(\mathbf{x}_{ij}, \rho_i, r_i^*\right) - \lambda_{ij} \cdot \kappa_j > 0\,, \\ 0, & \text{otherwise.} \end{cases} \tag{3}$$

Given the mode of computing hurdle rates and capital shares presented above, project realization is only attractive for the investment project yielding the highest NPV. Whenever projects are realized, departments are charged according to the relative benefit depreciation schedule (Rogerson 1997) using the hurdle rate as discount factor, and rewarded a function of residual income, $f(v_{it})$.[3] Representing the sequence of future compensation components by $\mathbf{v}_i = [f(v_{i1}) \ \ldots \ f(v_{iT})]$, we can denote department i's utility function by $U_i(\mathbf{r}(r_c \circ \mathbf{v}_i)$. Fulfilling the claim of

[2]This is the core of the coordination mechanism: the capital charge rate is located below the project's internal rate of return only for the winning project. For all other projects, the hurdle rate is higher than the projects' internal rates of return, which—upon realization—would result in negative NPVs (Baldenius et al. 2007; Leitner and Behrens 2013, 2014).

[3]In every t, residual income results in $v_{ijt} = x_{ijt} \cdot \rho_i - \lambda_{ij} \cdot \kappa_j \cdot \frac{x_{it}}{\mathbf{r}(r_i^*) \circ \mathbf{x}_i}$. This performance measure reduces to $v_{ijt} = \kappa_j \cdot \left(\rho_i - \rho_i^*\right)$ and, thus, fulfills strong incentive compatibility (cf. also Baldenius et al. 2007).

SHV maximization, we denote the corporation's utility function by $U_c(\Lambda_j(r_c, \mathbf{p}_j) - \mathbf{r}(r_c) \circ \mathbf{v}_i)$. Following the decision-making rule presented in Eq. (3) both utility functions are maximized.

3 Simulation Setup

We model department i's cash flow time series $\mathbf{x}_{ij}$ from operating project j by following a geometric brownian motion (GBM). In particular, we model two correlated cash flow paths (using a zero drift bivariate GBM) whereby one path corresponds to those actually put into effect and the second path is the department's forecast of cash flows. A process follows a zero drift bivariate GBM if it satisfies the stochastic differential equation:

$$dx_t^{(k)} = \sigma^{(k)} dB_t^{(k)} \tag{4}$$

where σ is the diffusion rate, B_t is a Wiener process, and $k = \{R, E\}$ denotes "real" and "estimated" values of cash flows. The Wiener processes are correlated such as that $\mathbf{E}\left[dB_t^{(R)} dB_t^{(E)}\right] = p_{R,E}dt$ and $p_{R,R} = p_{E,E} = 1$. We use a discrete time approximation[4] of the above process and (arbitrarily) set a constant $\sigma_{ij}^{(R)}$ for all i and j. Furthermore, x_{ijt} is normalized to one at $t = 0$. The diffusion rates of the real and estimated cash flow paths are linked according to $\sigma^{(E)} = \sigma^{(R)}/p_{R,E}$, with $p \in (0, 1]$. Hence, $p_{R,E}$ can be interpreted as a measure of each department's precision concerning estimating future cash flows. If $p_{R,E} = 1$, any two simulated trajectories of the bivariate GBM will perfectly coincide. $p_{R,E} < 1$ indicates uncertainty, estimated cash flows will then deviate from realized cash flows.

Paired trajectories of this process for $t = 1, 2, \ldots, T$ serve as the real and estimated cash flow time series for department i operating project j, respectively. The real efficiency parameter $\rho^{(R)}$ is drawn from a uniform distribution $\rho_{ij}^{(R)} \sim U(\underline{\rho}^{(R)}, \bar{\rho}^{(R)})$, however, department i faces uncertainty concerning its efficiency according to $\rho_{ij}^{(E)} = \min\left[\rho_{ij}^{(R)} \cdot \exp(y_{ij} \cdot \frac{1-p_{R,E}}{p_{R,E}} - \sigma_y^2/2), 1\right]$. y denotes a normal random variable with $y \sim \mathbf{N}(0, \sigma_y^2)$. The PV of each cash flow time series (real and estimated)—given the corporation's cost of capital r_c—is thus $PV_{ij}^{(k)}(\mathbf{x}_{ij}^{(k)}, \rho_{ij}^{(k)}, r_c)$. Each project's NPV is defined as $\Lambda_j^{(k)} = \sum PV_{ij}^{(k)}(\mathbf{x}_{ij}^{(k)}, \rho_{ij}^{(k)}, r_c) - \kappa_j^{(k)}$, and $\kappa_j^{(R)} \sim U(\sum PV_{ij}^{(R)} - \min(PV_{ij}^{(R)}), \sum PV_{ij}^{(R)})$. The latter definition of $\kappa_j^{(R)}$ guarantees non-negative NPV projects only and requires that every department i participates in project j. The estimated outlay $\kappa_j^{(E)}$ of project j is given by $\kappa_j^{(E)} = \kappa_j^{(R)} \cdot \exp(y_j \cdot \frac{1-p_{R,E}}{p_{R,E}} - \sigma_y^2/2)$.

[4]Using Matlab©'s function 'portsim.m'.

4 Results

This section presents results from a large variety of parameter sets. All relevant parameters either fixed or subject to variation in our simulation runs are presented in Table 1.

4.1 Erroneous Forecasts

We start by analyzing the effects of different corporate structures, i.e. the effect of different numbers of projects and departments operating these projects, respectively. Figure 1 reports results from 50,000 simulation runs concerning the ratio of erroneously chosen projects (loss ratio, LR). Light (dark) areas in each subplot indicate high (low) LR, subplots refer to values of $p_{R,E}$ of $\{.2, .4, .6, .8\}$ respectively. The abscissa (ordinate) shows varying numbers of departments (projects).

It turns out that LR is increasing in the number of departments as well as in the number of projects. In other words, "large companies" (either in terms of a large number of projects and/or a large number of departments) face a higher probability of choosing the wrong project due to inaccurate forecasts of cash flows efficiencies, as well as erroneous estimates of the project outlay. This is particularly true for setups with a relatively low forecasting ability (i.e., $p_{R,E} = .2$). Unsurprisingly, increasing forecasting abilities generally reduce LR. When looking at the NPV loss (NL)—which is the loss suffered from erroneously not picking the best investment project—we find that NL increases with the number of projects, but decreases as the number of departments increases (cf. Fig. 2). Further, the decreasing pattern

Table 1 Parameters in the simulation setup

Parameter	Description	Range
$\sigma_{ij}^{(R)}$	Diffusion parameter of the cash flow process	Fixed, $\sqrt{.1}$
$p_{R,E}$	Correlation of real and estimated cash flow trajectories	Variable, $\{.2, .4, .6, .8\}$, equal for all departments
T	Number of periods of the project	Fixed, 5
$\rho_{ij}^{(R)}$	Efficiency parameter of department i Operating project j	Random, $\rho_{ij}^{(R)} \sim \mathrm{U}(.1, .9)$
σ_y^2	Variance of the error in efficiency and outlay	Fixed, .1
r_c	Corporate cost of capital	Fixed, .1
$\kappa_j^{(R)}$	Initial outlay of project j	Random, uniform in the interval of positive NPV projects
n	Number of projects	Variable, $\{2, 3, 4, 5, 6, 7, 8\}$
z	Number of departments operating one shared project	Variable, $\{2, 3, 4, 5, 6, 7, 8\}$
s	Number of simulation runs	Fixed, 50,000

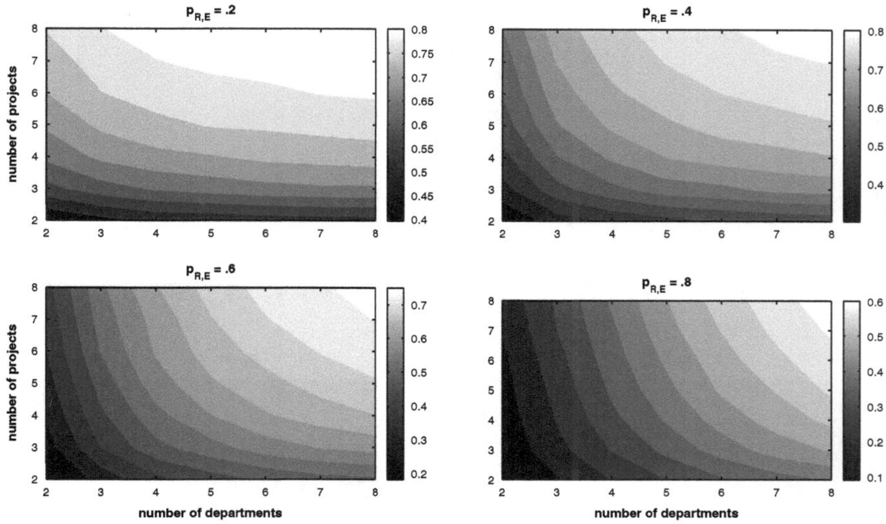

Fig. 1 LR for different combinations of n, z and $p_{R,E}$

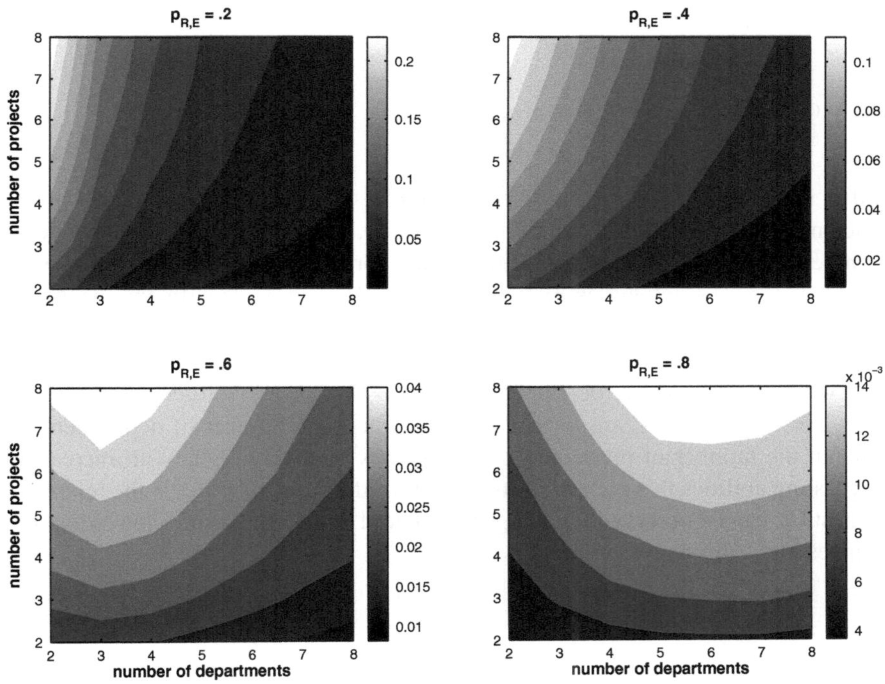

Fig. 2 NL for different combinations of n, z and $p_{R,E}$

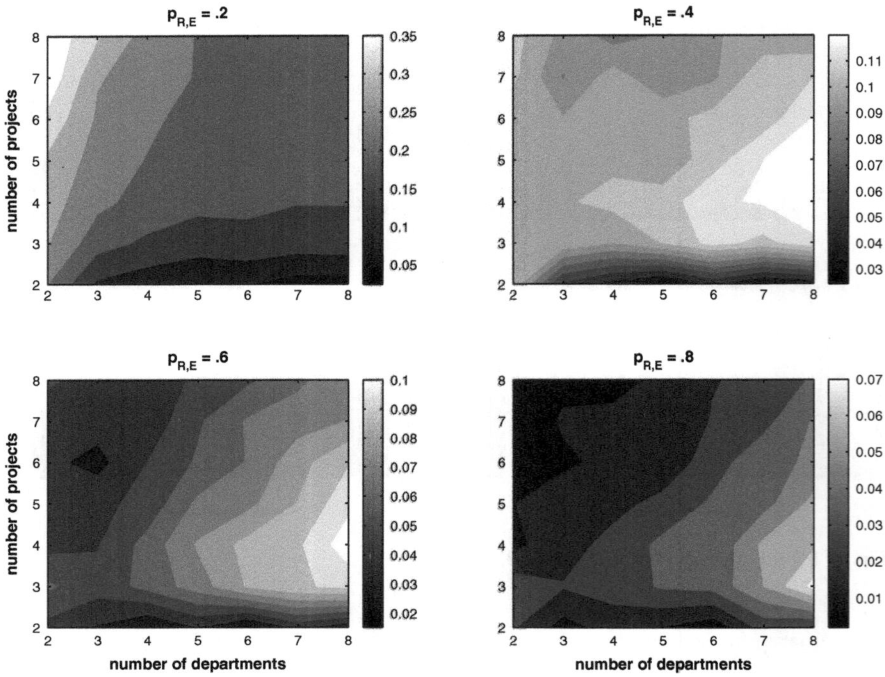

Fig. 3 CL for different combinations of n, z and $p_{R,E}$

shifts, if departments possess higher abilities in correctly forecasting cash flows, efficiency and outlay. For example, if $p_{R,E} = 0.8$, the NL is basically low, yet it is increasing with the number of departments. For the case of compensation losses (CL) (cf. Fig. 3)—defined as rewards[5] wrongfully paid to departments operating an adverse project due to errors in forecasts—we find that the highest errors can be observed for low forecasting abilities, i.e., for $p_{R,E} = 0.2$ the maximum is at 0.35 while for $p_{R,E}$ the errors upper boundary is 0.07. For low forecasting abilities ($p_{R,E} = 0.2$), high CL are located in areas where the number of departments is low but the number of investment alternatives is relatively high. With increasing forecasting abilities this pattern shifts so that for the case $p_{R,E} = 0.8$ the relatively highest CL can be observed for a larger number of departments and a lower number of investment alternatives.

[5]Based on the performance measure v_{it}, see Sect. 2.

4.2 *The Impact of Error Sources on the Mechanisms' Efficiency*

This section outlines effects of a reduced number of erroneous departmental estimates. Recall that in the basic setup cash flows **x**, outlays κ and the departments' efficiency parameter ρ are subject to misestimation. We provide results if uncertainty for one or two out of these three error sources is dropped. In other words, either one or two values from the set $\{x, \kappa, \rho\}$ is known at the very beginning of the project. We use the same parameter sets as before and conduct an analysis concerning LR, NL and CL, respectively.

For the sake of simplicity, we report all results graphically (captions of subplots refer to known parameters respectively) in terms of changes in error measures as compared to the standard setup where $\{x, \kappa, \rho\}$ are subject to all errors at the same time (i.e. Figs. 1 and 2). We further only report results on good and bad forecasting abilities ($p = \{.2, .8\}$).

Figures 4 and 5 depict the reduction of NL for different combinations of the number of departments and projects dependent on which parameter(s) can be perfectly forecasted by departments (cf. also Table 2). Obviously, for low forecasting abilities ($p = .2$), NL may be particularly lowered if the number of departments operating one joint project is low. This result applies to all versions of known parameters in the estimation process. However, when examining Fig. 5, it turns out that for high forecasting abilities perfect knowledge of one or two (out of three)

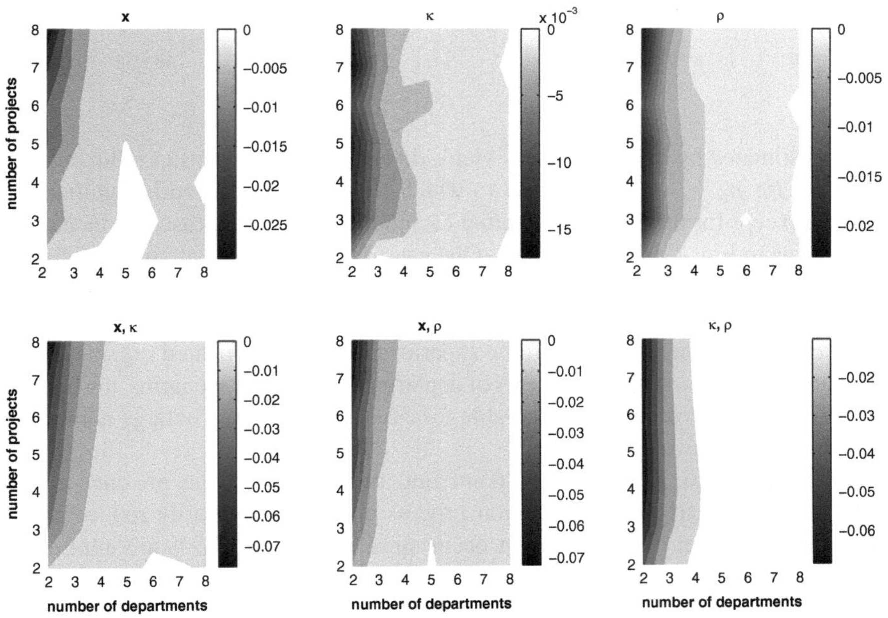

Fig. 4 Differences in errors for NL and $p = 0.2$

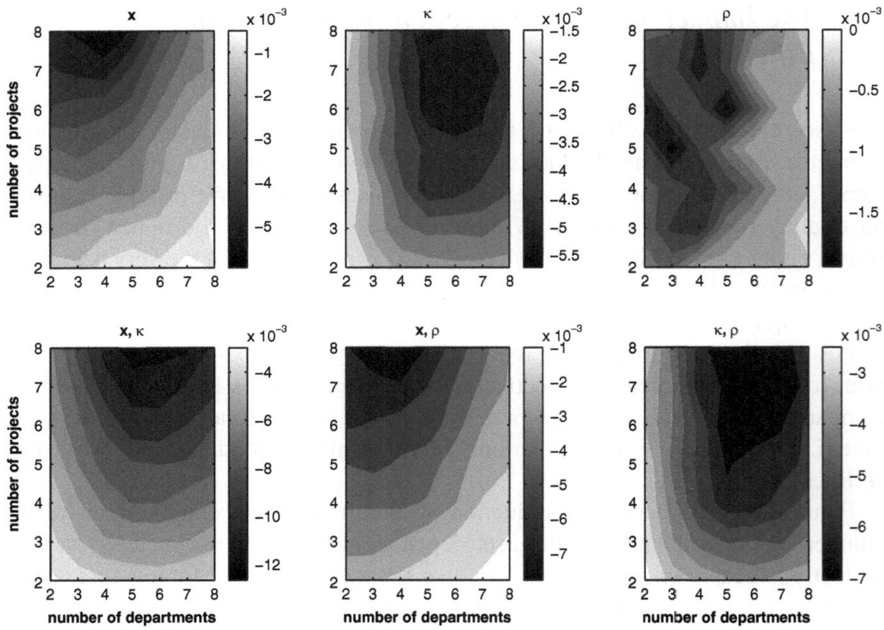

Fig. 5 Differences in errors for NL and $p = 0.8$

Table 2 Known/unknown parameters in alternative simulation setups

Setup	baseline	1	2	3	4	5	6
Known	–	$\{x\}$	$\{\kappa\}$	$\{\rho\}$	$\{x, \kappa\}$	$\{x, \rho\}$	$\{\kappa, \rho\}$
Unknown	$\{x, \kappa, \rho\}$	$\{\kappa, \rho\}$	$\{x, \rho\}$	$\{x, \kappa\}$	$\{\rho\}$	$\{\kappa\}$	$\{x\}$

values estimated by the departments yields diverse effects in terms of reducing NL. For the case $p_{R,E} = 0.8$ the extent to which NL can be improved is significantly higher (except for the case of the initial cash outlay, κ). In this case, no such clear patterns as in Fig. 4 can be observed. One can try to group the patterns as follows: for cases in which the initial cash outlay, κ, is known, organizations with a larger number of departments appear to be better off. For the remaining scenarios a lower number of departments appears to be superior to very sophisticated organizational structures (in terms of a large number of departments). In most scenarios, the highest potential for improvement is observable for cases with a relatively large number of investment alternatives

Interestingly, as Figs. 6 and 7 point out, perfect knowledge on one or two values in the departments' estimation process does not necessarily reduce errors in terms of a lower CL. This might occur since mitigating effects are eliminated too. In particular, for scenarios in which departments have a low forecasting ability ($p = 0.2$), perfect foresight on the capital outlay, κ, substantially increases CL. Further, we notice that errors are particularly lowered if the number of departments

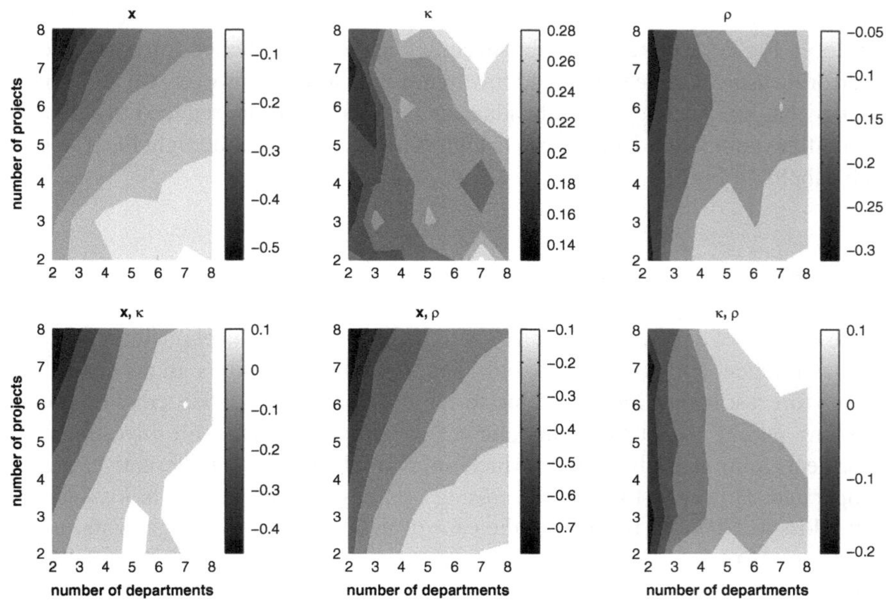

Fig. 6 Differences in errors for CL and $p = 0.2$

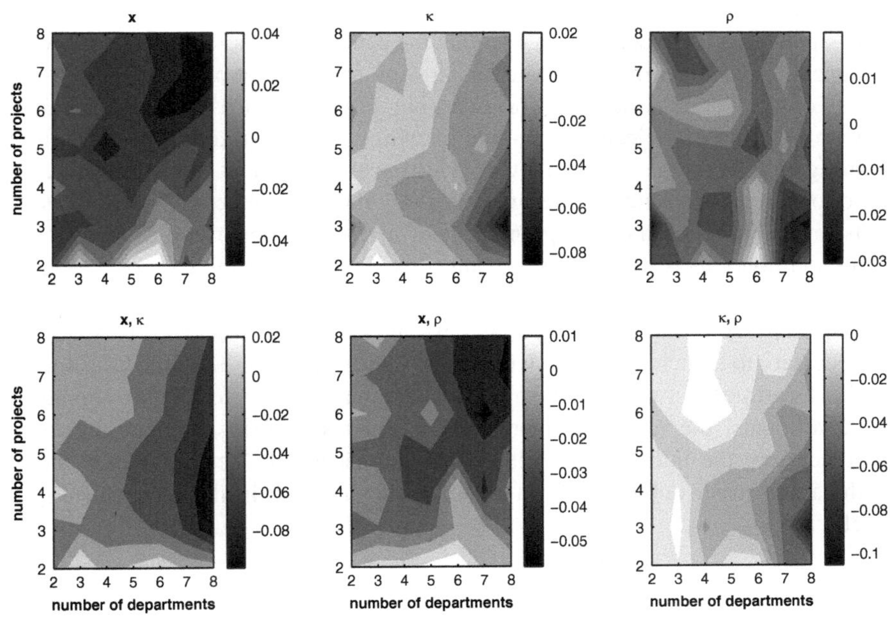

Fig. 7 Differences in errors for CL and $p = 0.8$

operating one joint project is low (see black areas to the left hand side of each subplot in Fig. 6). Contrary, for high accuracy in the departments' forecasts effects on error measures are scattered and low in absolute terms. However, we are able to identify that irrespective of which values are known, errors are reduced to a greater extent if the number of departments is high (see black areas to the right hand side of each subplot in Fig. 7).

Conclusion

Our results imply recommendations for the corporate structure (in terms of the number of departments jointly operating investment projects) with respect to an efficient coordination of distributed investment decisions. In the case of low forecasting abilities (e.g. due to hardly predictable market conditions), a low number of departments and a low number of available investment projects minimizes the ratio to which unfavorable investment alternatives are operated. However, for low forecasting abilities the compensation loss and the loss in NPV are reduced in the case of very sophisticated organizational structures (many departments) and (in most cases) a low number of available investment alternatives. This consequently indicates that, for the case of low forecasting abilities, organizations are better off if they are allowed for (extensive) cooperation in operating joint investment projects. If forecasting abilities are high with respect to the efficiency of the coordination mechanism designing the organization in a way that the extent of cooperation is kept low appears to be superior to unlimitedly allowing for cooperation.

For setups in which one or two (out of three) measures associated with investment projects are ex-ante known, we reveal a high potential for mitigating NL in the case of a low forecasting ability. This is particularly true for organizations which are designed in a way that the number of departments is low. Here, a relatively large number of investment opportunities leads to an additional increase in the mechanism's efficiency. For the case of a high forecasting ability, results are diverse—no clear pattern can be observed. With respect to CL, it has to be noted, that ex-ante knowing the amount of money necessary to launch the investment project significantly increases the basis for the departments' variable compensation component. Thus, for organizational departments, the mechanism provides incentives to focus particularly on forecasting the initial cash outlay. With respect to NL, better forecasts of the initial cash outlay result in a significant increase in the mechanism's efficiency, too.

References

Arya A, Glover J, Young RA (1996) Capital budgeting in a multidivisional firm. J Acc Audit Financ 11(4):519–533

Axtell RL (2007) What economic agents do: how cognition and interaction lead to emergence and complexity. Rev Austrian Econ 20(2–3):105–122

Baldenius T (2003) Delegated investment decisions and private benefits of control. Acc Rev 78(4):909–930

Baldenius T, Dutta S, Reichelstein S (2007) Cost allocation for capital budgeting decisions. Acc Rev 82(4):837–867

Dutta S, Fan Q (2009) Hurdle rates and project development efforts. Acc Rev 84(2):405–432

Georgiades S, Mavronicolas M, Spirakis P (2000) Optimal, distributed decision-making: the case of no communication. In: Fundamentals of computation theory. Springer, Berlin, Heidelberg, pp 293–303

Hendry J (2002) The principal's other problems: honest incompetence and the specification of objectives. Acad Manag Rev 27(1):98–113

Irlenbusch B (2006) Experimental perspectives on incentives in organisations. Cent Eur J Oper Res 14:1–24

Kirman AP (1993) The economy as an interactive system. In: Arthur WB, Durlauf SN, Lane DA (eds) The economy as an evolving complex system, vol 2. Addison-Wesley, Reading

Kouvelis P, Lariviere MA (2000) Decentralizing cross-functional decisions: coordination through internal markets. Manag Sci 46(8):1049–1058

Leitner S (2012) Information quality and management accounting. Lecture notes in economics and mathematical systems, vol. 664. Berlin, Heidelberg, New York: Springer

Leitner S (2012) Interactions among biases in costing systems: A simulation approach. In: Andrea T, Simone A, Eva C-G and Miguel G-V (eds) Managing market complexity, Berlin, Heidelberg, New York: Springer, pp. 209–220

Leitner S, Behrens DA (2013) On the fault (in)tolerance of coordination mechanisms for distributed investment decisions. Cent Eur J Oper Res. doi:10.1007/s10100-013-0333-4

Leitner S, Behrens DA (2013) Residual income measurement & the emergence of cooperations: results of an agent-based simulation. In: Leopold-Wildburger U, Dargam F, Pickl S, DeTombe D, Pia L, Liu S, Hernández J, Delibasic B, Ribeiro R, Zarate P (eds) Proceedings of the EURO-mini conference Graz 2013, p 26. EWG-E.CUBE, EWG-DSS, EWG-MCSP, EWG-ORAFM, Graz

Leitner S, Behrens DA (2014) On the robustness of coordination mechanisms involving incompetent agents. In: Leitner S, Wall F (eds) Artificial economics and self organization, vol 669. Springer, Berlin, pp 191–203

Leitner, S (2014) A simulation analysis of interactions among intended biases in costing systems and their effects on the accuracy of decision-influencing information. Cent Eur J Oper Res. 22(1):113–138

Rogerson WP (1997) Inter-temporal cost allocation and managerial investment incentives: a theory explaining the use of economic value added as a performance measure. J Polit Econ 105:770–795

Ryan PA, Ryan GP (2002) Capital budgeting practices of the fortune 1000: how have things changed? J Bus Manag 8(4):355–364

Schuster P, Clarke P (2010) Transfer prices: functions, types, and behavioral implications. Manag Acc Q 11(2):22–32

Young SD, O'Byrne SF (2002) EVA and value-based management. A practical guide to implementation, 6th edn. McGraw-Hill, New York

Why Do Firms Exist?

Vipin P. Veetil

1 Introduction

> An outsider to the field of economics would probably take it for granted that economists have a highly developed theory of the firm. After all, firms are the engines of growth of modern capitalistic economies, and so economists must surely have fairly sophisticated views of how they behave. In fact, little could be further from the truth (Hart 1989, p. 1757).

The basic economic problem is how to allocate resources in a system where information about preferences, endowments and production possibilities is dispersed among the many (Hayek 1945; Hurwicz 1973). In some areas of economic activity allocation happens through the "invisible hand" of markets, while in other areas one sees the visible hands of managers. What accounts for the difference in economic organization?

In a dispersed information environment, the problem of allocation involves *communication* and *computation*. Private information has to be conveyed to decision

This essay benefited immensely from suggestions by Peter Boettke, Christopher Coyne, Andrew Crooks, Virgil Storr, Vlad Tarko, Pietro Terna and members of the Graduate Student Paper Workshop at George Mason University. Thanks is also due to Surajit Mazumdar, for much of this essay is a response to a challenge he posed when the author was an undergraduate student at University of Delhi. All errors are the author's own.

V.P. Veetil (✉)
Department of Economics, George Mason University, Enterprise Hall, 3rd Floor, MSN 3G4, Fairfax, VA 22030-4444, USA
e-mail: vpudiyad@gmu.edu; vipin.veetil@gmail.com

F. Amblard et al. (eds.), *Advances in Artificial Economics*, Lecture Notes in Economics and Mathematical Systems 676, DOI 10.1007/978-3-319-09578-3_2

making agents, who in turn must compute allocations.[1] Markets and firms are different ways of "communicating widely dispersed information" (Myerson 2008, p. 586). Markets use decentralized communication where individuals convey private information to each other, often through a system of prices. Firms use centralized communication where private information flows from workers to managers and commands flow in the opposite direction. The difference in the architecture of interactions in markets and firms means that they perform differently on two margins. *First*, the way in which individual level noise affects system level performance. *Second*, the speed with which they allocate resources.

Like physical and biological systems, communication in economic systems involves noise, i.e. the passing of incorrect information. Noise arises because of both intentional and unintentional reasons. Economic agents intentionally transmit false information when private interests dictate them to do so. This is leads to *adverse selection* and *moral hazard* problems (Myerson 2008). The mechanism design literature points to a variety incentive-compatible rules to solve these problems. However in so far the rules are not perfectly tailored to the real world, there will be some noise in communication. Economic agents may unintentionally transmit false information because some kinds of information is difficult to encode accurately (Polanyi 1966). In firms information is aggregated and provided to the planner-manager. Local noise is aggregated and enters the planner-manager's decision making process. In other words, local noise has a direct global impact. And this lowers the efficiency of the allocation made by the planner-manager. In markets information is communicated in a decentralized manner. Local noise enters only local decision making, its impact on global allocation is more limited. Decentralization of communication lowers the impact of noise on resource allocation.

Though markets are more robust to noise, firms take less time to allocate resource. The absence of a central decision maker means that markets, unlike firms, have to muddle through numerous interactions between agents to allocate resources. There exists a trade-off between *speed* and *efficiency*. Markets are more efficient, but firms are faster. The parameters of this trade-off depend on a variety of factors including the nature of information communicated and the types of individuals communicating it. These parameters vary across different areas of economic activity. In some areas significant gains in efficiency may be made with a small reduction in speed, in other areas the converse may be true. This means that markets will be preferred in some areas and firms in others. The theory propounded in this essay explains a variety of empirical facts like why the largest barber shops tend to be smaller than the largest retail stores, and why markets tend to be replaced with planning during wars. It has implications for economic development, and can be extended to understand churches, governments and other non-market organizations.

[1] In this essay the problem of agent-level computation is assumed away. Each agent has access to an oracle that instantaneously computes the right answer to problems posed to it, given the requisite information. The oracle has no independent means of getting the requisite information.

An agent-based model is used to study the relative performance of a firm and a market as the amount of noise in communication varies. This essay merely sketches the broad contours of the theory. And the agent-based model is even simpler. Section 2 discusses related literature. Section 3 develops the theory and Sect. 4 provides the structure of the agent-based model. Section 5 lists the primary results generated by the model. These results are based on data gathered by running a parameter sweep. Section 6 is at once speculative and promising for it discuss future work. Section 7 offers concluding remarks. A pseudo-code is provided in section "Concluding Remarks".

2 Related Literature

Why do firms exist? The question relates to two areas of economics: theory of firm and mechanism design.

2.1 Theory of the Firm

Coase (1937) asked "why co-ordination is the work of the price mechanism in one case and of the entrepreneur in another" (p. 389). Coase was of the view that the use of the market mechanism involves costs such as finding trading partners, negotiating price and quantity et al. And that these costs can be avoided by allowing a planner-manager to make decisions. However as a firm grow larger inefficiencies set in and this limits its size. In other words, as to whether an activity is to be done within a firm or through markets is determined by the marginal cost of using the two mechanisms.

Coase's paper languished in the backwaters of economic theory for nearly three decades before a revival began in the 1970s. The renewed interest in the subject in the 1970s lead to a whole host of theories of why firms exist. These can be classified into six groups: transaction cost, team production, liquidity preference, assess-specificity, and nexus-of-contracts.[2]

Coase (1937) proposed the *transaction cost* theory of the firm. Which says that whether an activity is done within a firm or through markets depends on the marginal cost of the two mechanisms. Coase (1988, p. 40) however "did not attempt to uncover the factors that would determine" the source of the difference in relative costs. In essence Coase asked the question but did not provide an answer.

Alchian and Demsetz (1972) proposed the *team production* view. Alchin and Demsetz began with the postulate that team production has the potential to increase output (basically the idea that specialization increases productivity). However if

[2]For a literature review of theories of firm see Hart (1989); the classification in this essay is mildly different from that of Hart (1989).

each member of the team is not rewarded in accordance with her contribution to total output, she has an incentive to shirk. This means that teams that develop methods to observe and meter the contribution of different members will be more productive. One way to reduce shirking is to appoint a planner-manager to oversee team production. The planner-manager's incentives can be aligned by making her the claimant of residual income. And this creates a firm.[3] The team production view however confuses the problem with the solution. Markets too involve team production, in the fact the whole economy can be thought as one big team which benefits from specialization and exchange. Markets monitor team production using a system of profit and loss. The problem of monitoring team production in a firm arises after the decision is made to conduct an activity within a firm and not through markets. The appointment of a manager solves a problem that is created because of the use of a non-market mechanism. It does not explain the use of the non-market mechanism in the first place.

Simon (1951) proposed the *liquidity preference* view of the firm. The basic idea is that a preference for employment contracts over other contracts arises out of the fact that employers do not know what exactly they would want employees to do in the future. And faced with this uncertainty employers would like a contract that allows them to be in a command and control position. In so far as employees are indifferent between a broad range of tasks, employment contracts provide gains to both parties. However the best means to postpone uncertainty is not to sign a contract at all. Money is the ultimate form of liquidity.

Williamson (1975) proposed the *asset-specificity* view. The basic idea is that capital equipment in modern industries tends to be tailor-made to perform specific tasks. And this makes the owners of such assets vulnerable to opportunistic behavior by the buyers of the assets' services. One way to solve this problem is vertical integration, i.e. take the activity that would have otherwise been conducted through the market into the purview of the firm. Williamson assumes that agents cannot overcome the threat of opportunistic behavior by writing complete contracts because they have bounded rationality.

There are three shortcomings of the asset-specificity view. *First*, presumably there is a cost to vertical integration, for otherwise the whole economy would integrate into a single firm. Yet the view does not explain the origin or existence of these costs. *Second*, it assumes—albeit implicitly—that there cannot be cross-ownership across firms. *Third*, the folk-theorem says that in infinitely repeated games a variety of cooperative equilibrium, which were not possible in finite games, can be realized. While a firm may benefit from exploiting another firm in one game, it will pay the cost of losing reputation and therefore future business possibilities. Reputation is a substitute for vertical integration, the asset-specificity view does not say which will be used when and why.

[3] **?** develops an agent-based model of the team production idea and reproduces numerous stylized facts about US firm size distribution.

Jensen and Meckling (1976) proposed the *nexus-of-contract* view. It says that relations between individuals in a firm is no different from relations between individuals in a market. In both cases economic agents have contractual arrangements with each other, the question of "why firms exist" is misplaced. This view overlooks an important difference between contracts in firms and in markets.

> At this stage, it is important to note the character of the contract into which a factor enters that is employed within a firm. The contract is one whereby the factor, for a certain remuneration (which may be fixed or fluctuating), agrees to obey the directions of an entrepreneur within certain limits. The essence of the contract is that it should only state the limits to the powers of the entrepreneur. Within these limits, he can therefore direct the other factors of production (Coase 1937, p. 391).

Compare this to market contracts that specify price and quantity of exchange. While contracts within firms specify "negative rights" in the sense that managers may not ask workers to do certain activities, market contracts specify "positive rights" in the sense that parties have an obligation to perform certain activities. The question of why firms exist is essentially a question of why these different contracts exist. The nexus-of-contract view fails to appreciate the question. There has been little innovation in the theory of firm since the late 1980s. The question of "why firms exist" remains open.

2.2 *Mechanism Design*

Hayek (1945) laid out a challenge for the economics profession. Hayek said that the basic economic problem is how to best use information which is dispersed among the many. This meant finding solutions to two problems. *First*, how to communicate information of "the kind which by its nature cannot enter into statistics and therefore be conveyed to any central authority in statistical form" (Hayek 1945, p. 524)? *Second*, how to communicate information fast enough? Hayek's seminal paper however was more thoughts than theorems.

Hayek's challenge was taken up by Marschak (1959), Hurwicz (1973) and others. And this gave birth to the field of mechanism design. Work in this area compared the relative performance of market and non-market mechanisms, and the rules under which individuals have an incentive to reveal private information to one another. However the mechanism design literature has yet to answer the questions Hayek posed in 1945 for two reasons. *First*, the literature works with information that can be easily encoded and transmitted without noise. There is no noise if the incentive compatibility constraint is meet. Sah and Stiglitz (1987) is an exception. *Second*, the literature does not study the time it takes to communicate information.[4] Bolton and Farrell (1990) is an exception.

Sah and Stiglitz (1987) study the impact of noise in communication but do not study the time it takes for different systems to allocate resources. Bolton and Farrell (1990) study time but do not allow agents to communicate with each other.

3 The Theory

Firms and markets are two different mechanisms of allocating resources. In a firm dispersed information is first brought to a central authority, say a planner-manager. The manager then computes efficient allocation of resources and conveys instructions to everyone. In a market dispersed information is never brought to a central authority, resources are allocated through the interaction between agents (Axtell 2003). In the process of market interactions agents communicate bits of information to each other, but no agent come to hold all information.

In a system with n agents how many interactions will be necessary to allocate resources? In a firm the planner-manager meets each agent once to collect information and then makes allocations. Therefore n interactions are necessary. As to how many interaction are necessary in a market depends on a variety of factors including the distribution of dispersed information and the complexity of the allocation problem.[5] Assume that more interactions are necessary in markets than in firms. If for instance, each agent in the market must meet every other agent once, then $\frac{n(n-1)}{2}$ interactions will be needed. If a system with 100 agents, a market will need nearly 50 times as many interactions as a firm. The cost of decentralized communication is an increase in the number of interactions.

Why do markets exist if firms can allocate resources with fewer interactions? So far it was assumed that when two agents meet they convey information to each other with perfect accuracy, i.e. the probability of error is zero. Suppose there exists a small probability that an agent will commit an error when conveying information to another agent. This is true both when agents meet each other in a market and when they meet a planner-manager in a firm. The error introduces noise in communication. In the presence of noise both firms and markets perform worse

[4] As to why much of the literature ignores this problem may have to do with the use of formal logic where "the only thing that is important is whether a result can be achieved in a finite number of elementary steps or not. The size of the number of steps which are required, on the other hand, is hardly ever a concern of formal logic. Any finite sequence of correct steps is, as a matter of principle, as good as any other. It is a matter of no consequence whether the number is small or large, or even so large that it couldn't possibly be carried out in a lifetime, or in the presumptive lifetime of the stellar universe as we know it" (Von Neumann 1951, p. 15).

[5] Needless to say, no interactions are necessary if market participants have access to equilibrium prices. However equilibrium themselves have to be discovered, and there is no way to do this except through interactions between economic agents.

than before in reaching efficient allocation. However firms are affected more than markets. This is because firms aggregate noise with information. And therefore local noise has a direct impact on global allocation. Whereas in a market local noise affects local allocation, its impact on global allocation is limited and indirect.

In a system with a small probability of error when one agent passes information to another, there exists a trade-off between speed and efficiency in the allocation of resources. Firms may allocate resources faster but will be less efficient that markets. The tradeoff varies across economic activities because of differences in the nature of information to be conveyed and the individuals conveying them. The existence of a speed-efficiency tradeoff and its variation across activities means that the optimal form of economic organization is unlikely to be a corner solution, i.e. there is reason why both firms and markets exist.

The following analogy from the human nervous system is useful in understanding the difference between firms and markets. Imagine placing your hand in a tub of water at 80 °C. There are two ways to convey this information from the nerve endings to the brain. *First*, convert the temperature to binary form (101000) and then transmit six bits of information. *Second*, transmit 80 bits, with one bit for each 1 °C. Interestingly, human nervous system uses a mechanism akin to the second method despite the fact that it takes more than thirteen times as much time as the first method.[6] The reason is that the second method is more robust.

> ...the counting methods has a high stability and safety from error. If you express a number of the order of a million by counting and miss a count, the result is only irrelevantly changed. If you express it by (decimal or binary) expansion, a single error in a single digit may vitiate the entire result (Von Neumann 1951, p. 101).

3.1 Three Questions

This section addresses three important questions. *First*, what is the relation between time and number of interactions? *Second*, what kinds of architectures can firms and markets use to reduce the impact of noise? *Third*, where does noise come from?

What is the relation between time and number of interactions? Both in markets and firms, multiple interactions can happen simultaneously in a given time period. In a market several people may trade simultaneously. In a firm several lower-level managers may meet several workers simultaneously, and then convey this information to a higher-level manager. More generally, the price system in markets and communication systems in firms are means by which the two mechanisms reduce time it takes to convey information. Let $f(i)$ and $m(i)$ represent the time i interactions take in a firm and a market respectively. Suppose $g(i)$ interactions are necessary in a market for every i interactions in a firm. Earlier it was said that $g(i) > i$ is a necessary condition for the theory propounded in this essay. If the

[6]This is assuming that coding and decoding of messages take the same time in both cases.

architecture of markets and firms is such that multiple interactions can happen in the same time period, then the necessary condition must be modified to $m \circ g(i) > f(i)$.

What kinds of architectures can firms and markets use to reduce the impact of noise? Biological and physical systems employ a variety of architectures to reduce the impact of noise and failure of component parts. Redundancy and duplication are well-known engineering tricks. Markets may be structured so that agents have an incentive to meet multiple times, firms may require workers to meet managers frequently. In firms the impact of noise may be reduced by building hierarchical structures where some component sub-parts are similar. Simon's (1962) watch makers Hora and Tempus illustrate the use of hierarchical structures in a system where component parts fail with a small probability.

Where does noise come from? There are two distinct sources of noise in communication. The first source of noise is the nature of economic agents, "individuals will not share private information or exert hidden efforts without appropriate incentives" (Myerson 2008, p. 587). Individuals may intentionally convey incorrect information if it is in their interest to do so. The second source of noise is the nature of economically relevant information. It is conceivable that some information may be difficult to encode accurately (Polanyi 1966), such information has come to be known as 'tacit knowledge'.

4 The Model

4.1 Why an Agent-Based Model?

Agent-based models are useful to solve problems for which it is difficult to write down equations that fully describe the behavior of the system (Axtell 2000). Writing down equations becomes increasingly difficult as one begins to incorporate realistic assumptions about economic agents like bounded rationality and decision making under local information. While there do exist analytical models of bounded rationality (Rubinstein 1998), they say very little about out-of-equilibrium dynamics. In some economic problems, like the impact of a tax on consumption of cigarettes, out-of-equilibrium dynamics may not be of the greatest interest. While in other problems, like firm formation and business cycles, out-of-equilibrium dynamics constitute their very essence.

4.2 Model Setup

A market and a firm are compared. Both have n agents. For the purposes of this essay n is set to 100. Agents have a utility function defined over two goods and receive an initial endowment. The initial endowment is not a Pareto optimal allocation, the problem is how to reallocate goods.

The firm and the market are two ways to reallocate goods. In the market agents meet through a process of binary matching, exchange ratio is fixed at one. Every binary match two agents are allowed to trade one unit of a good for one unit of the another good, if both wish to trade. In the firm the planner-manger collects information about each agent's endowment and then reallocates goods so as to maximize a social welfare function (SWF).

Agents make errors while reporting their endowment both in the market and in the firm. The probability of error is a parameter. This means that in the market agents may engage in trades that make them worse-off and in the firm agents' errors may cause the planner-manager to misallocate goods.

The performance of the economy is given by the distribution of utilities of all the agents. The distribution is mapped to the real number line using a SWF. A mapping from a distribution to the real line necessarily means loss of information. The choice is really as to what kind of information to retain and what to let go. The SWF is defined as the product of the utilities of all agents. This function captures two ideas. *One*, for given second and higher moments, the SWF is monotonically increasing with respect to the first moment. *Two*, for a given first moment, the SWF is monotonically decreasing with respect to the second moment. The SWF captures the idea of complementarity between the well-being of different agents. Though the model is that of an endowment economy, the two properties of the SWF get to some of the consequences of the cobweb of interrelations that define a system of industrial production.

Several factors are left out because of the belief that "good fences make good models". The market does not have prices, and the firm does not have communication technology to aggregate information. The firm does not have layers of hierarchy. The market and the firm do not interact. Plans to incorporate these features are discussed in Sect. 6.

4.3 Agents

Each agent is endowment with two goods: x_a and x_b . Agents maximize a utility function subject to a constraint. e_a and e_b denote the initial endowment of the two goods. The second good is the numeraire.

$$\max_{x_a, x_b} U(x_a, x_b) = \sqrt{x_a} + \sqrt{x_b} \tag{1}$$

$$\text{subject to} \quad p_a x_a + x_b \le p_a e_a + e_b \tag{2}$$

Initially an agent is endowment with either *two units of* x_a *and zero units of* x_b or *zero units of* x_a *and two units of* x_b, i.e. $\{e_a, e_b\}$ is either $\{0,2\}$ or $\{2,0\}$. Substituting endowments into demand function and setting price equal to 1 yields demands.

$$x_a^* = x_b^* = 1 \tag{3}$$

In the market each agent's problem is to go from the initial endowment to optimum consumption. In the firm, the planner-manager's problem is to take each agent from initial endowment to optimum consumption because of the way the SWF is defined. Therefore, the market and the firm try to maximize the same function but in different ways.

4.3.1 Attributes

Each agent has the following data storage objects.

1. Endowment_list $[e_a, e_b]$ stores an agent's initial endowment.
2. Goods_list $[x_a, x_b]$ stores the amount of goods an agent has at a point in time.
3. Utility_list records the utility an agent receives if it consumes the goods it has.

Each agent has the following functions or capabilities.

1. trade_decision(a,b) takes two parameters: a and b. The function asks the agent whether it wants to give one unit of good b and receive one unit of good a. The agent compares the quantities of a and b in the Goods_list. The function returns "True" if and only if the quantity of a is less than quantity of b.
2. report_goods(*noise*) takes one parameter: *noise*. It asks the agent to report the quantity of goods it currently has. The function returns a list with the quantity of x_a as the first element and x_b as the second element. The *noise* parameter defines the probability of making an error while reporting. If $noise = 0.1$, then with 10 % probability the agent will make an error in reporting the quantity of x_0 and x_1. If the agent makes an error, then it is equally likely to report a quantity one unit greater or one unit lower than the actual quantity.
3. report_endowment() does not take any parameters. It asks the agent to report the initial endowment of goods. The function returns the initial endowment.
4. compute_utility() does not take any parameters. It asks the agent to compute its utility from consumption of the goods it currently has. The agent plugs in the quantities of the goods it has into the utility function and returns the resulting value.
5. result() does not take any parameters. It asks the agent to update the Goods_list after a trade is completed. The agent reduces the quantity of the good it gave by one unit and increases the quantity of the good it received by one unit.

4.4 *Market*

1. Create a list of agents
2. Randomly sample two agents from the list
3. Ask each agent if it wants to trade using *trade_decision*() function
4. If both want to trade, let trade happen and ask agents to update the goods they have using the *result*() function
5. Compute and record the SWF using the *utility_markets*() function. The *utility_markets*() function returns the product of utility of all agents.
6. Repeat the process *n* times, in each round there are as many binary matches as number of agents.

4.5 *Firm*

1. Create a list of agents.
2. Ask each agent how much of each good it has, using the *report_goods*() function.
3. Give and take goods from agents. If an agent says it has two units of a good then take one unit away, if an agent says it has zero units of a good then give one unit. And if an agent has one unit of both goods, then do nothing.
4. Compute and record the SWF, using *utility_planner*() function, it returns the product of utility of all agents.
5. Repeat the process *n* times, i.e. once for every agent.

4.6 *Parameters*

The model has two parameters: *noise* and *number of agents*. For the purposes of this essay number of agents is set to 100. Theoretically noise can vary from zero to one, however the parameter sweep tests for the performance of the system for noise from 0 to 0.05, beyond which both the market and the firm degenerate.

Both the market and the firm play the game for 500 periods. The game begins with the creation of a list of agents and initial endowments of goods. The market and the firm have separate lists of agents. Every period in the market as many binary matches are created as the number of agents, in the firm the planner-manager meets each agent to receive information on the goods the agents currently have and then reallocates goods. And the process continuous for a 500 periods. The value of the SWF is recorded at the end of every period. However in the market agents are randomly selected and therefore not all agents may get a chance to trade every period. In the firm, each agent meets the planner once every period. The market follow random activation whereas the firm follows uniform activation.

A parameter sweep is run by playing a game with 500 periods for every parameter value. The noise parameter is ranged between 0 and 0.05 with increments of 0.001. This generated a total of 25,000 data points.

4.7 Verification and Validation

The purpose of *validation* is to test whether a model "corresponds to the real world system it is supposed to represent". The toy model presented in this essay is barely at stage zero on the Epstein-Axtell scale of agent-based models. It produces the qualitative idea that firms perform better than markets under certain conditions and vice-versa, this means that there is reason for both firms and markets to exist. The model is however far too primitive for validation.

The purpose of *verification* is to test whether a model "does what it is supposed to do", i.e. to check if the model is consistent with the theory that it implements. There are number of ways to verify a model. *First*, compare the results of the model to what the theory predicts. Von Neumann (1951) says that the impact of noise on system of communication varies according to how information is communicated. This is found to be true. Noise, i.e. errors by agents in communicating information on goods they hold, has more of an impact on firms than markets. *Second*, compare the results of the model to that of other similar models. This cannot be done because no other theory of firm incorporates noise and time. *Third*, study the evolution of the system overtime. Graphics on the behavior of the firm and the market overtime are produced. Their behavior is consistent with what the theory predicts. In a system without noise firms reach maximum social welfare in the first round, whereas markets reach maximum over several rounds. *Fourth*, trace the evolution of a single agent to see if its dynamics is consistent what is expected. Several individual agents were studied, particularly at the debugging stage. In a system without noise, over time agents tend towards having one unit of each good. After the introduction of noise, it was found that agents make the error of over reporting and under reporting by one unit. This is what was expected. *Fifth*, test the internal validity of a model by comparing outcomes over several runs. Outcomes of several simulation runs were compared to spot any red herrings. This included spanning the parameter space over noise and number of agents. As noise increases, the performance of both firms and markets worsens.

5 Results

Hypothesis 1 *In a system with zero noise the market attains maximum social welfare. It takes 27 periods.*

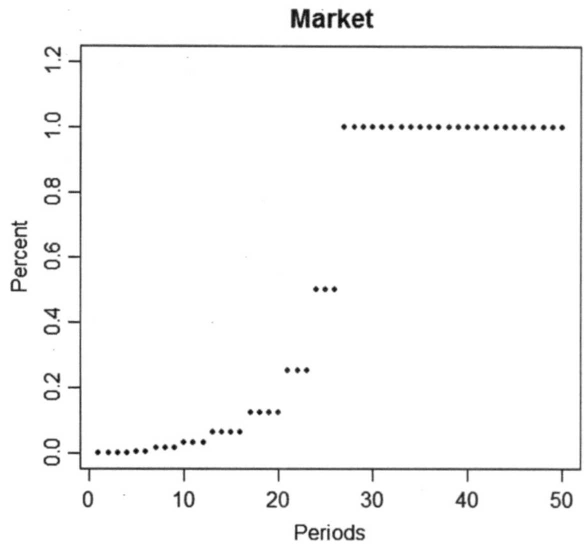

Fig. 1 Market with zero noise

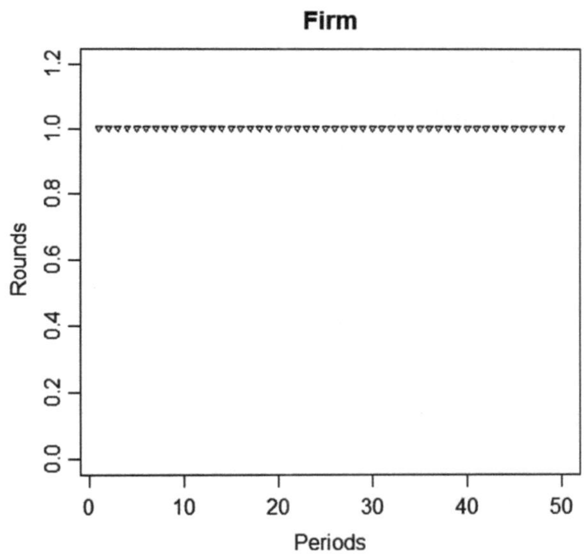

Fig. 2 Firm with zero noise

Evidence and Discussion See Fig. 1. The y-axis shows the proportion of maximum social welfare the system attains, the x-axis shows the number of periods. The market attains maximum social welfare in 27 periods, i.e. after 2,700 binary interactions between agents.

Hypothesis 2 *In a system with zero noise the firm attains maximum social welfare. It takes one period. The firm takes less time than the market.*

Evidence and Discussion See Fig. 2. Unlike the market, the firm reaches maximum social welfare in the very first period, i.e. after 100 interactions between agents and the planner-manager.

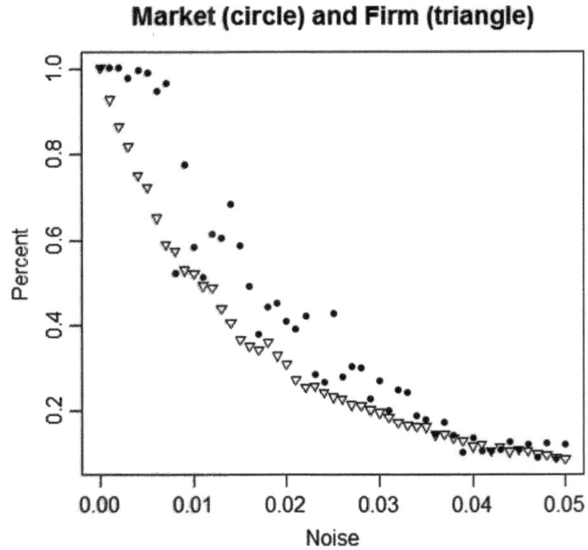

Fig. 3 Performance of the market and the firm in the presence of noise

Hypothesis 3 *The market is more efficient than the firm in the presence of noise.*

Evidence and Discussion See Fig. 3. The figure plots the average social welfare attained by the market and the firm in the last 400 rounds over different levels of noise. The first 100 rounds are disregarded so as not confound questions of efficiency and speed. Triangles represent the firm, circles represent the market. Without noise the circle and the triangle are on top of each other, i.e. both the firm and the market reach same level of efficiency. With noise the circles are largely above the triangles, i.e. the market reaches a higher efficiency than the firm.

6 Implications

The theory presented in this paper has several implications. *First*, as communication technology improves—ceteris paribus—there will be larger firms. The accuracy and cost of communicating information is a determinant of the trade-off between speed and efficiency. More accurate ways of communicating information will allow firms to reach greater efficiency. This explains the emergence of behemoths like Walmart over the last few decades. Never before has a single firm employed nearly one percent of the US labor force. Advances in communication technology is a driver of the Walmart phenomena.

Second, why are the largest barbershops smaller than the largest retails stores? Certain personal services like massages and haircuts involve information that is more difficult to convey accurately to higher levels of managements, i.e. they involve more "tacit" knowledge (Polanyi 1966). Retails giants on the other hand convey information on more standardized products. The variation in firm size by product type is not explained by any other theory of the firm.

Third, why do markets live longer than firms?[7] A great many firms come to live and die every year, yet market economies tend to last for centuries. Some firms too last for centuries, however the distribution of life of markets probably has a higher first moment than the distribution of life of firms. Markets live longer than firms simply because markets are more robust to errors.

Fourth, how does economic development depend on economic organization? The difference between rich and poor nations "lies largely in matters of economic organization" (Stiglitz 1989, p. 202). The optimal mix of firm and markets will depend on a variety of factors including dispersion of information, the nature of information, technology et al. Changes in economic fundamentals will alter the trade-off between speed and efficiency in different areas, and will change the optimal form of economic organization. Systems that have an incentive to find the new optimal organization in response change will tend to fare better than systems that do not have such incentives.

Fifth, what determines the optimal size of governments, churches and families? Perhaps a speed-efficiency trade-off akin to the tradeoff that determines the optimal size of firms.

Sixth, why is there redundancy in the way information is collected within firms? Feldman and March (1981) find that decisions makers in firms collect more information than needed to make decisions. And, information is collected in "surveillance mode" rather than decision making mode. They propose a psychological and sociological explanation of this "inefficient behavior". There is a simpler economic explanation. It is well know that redundancy is useful for reducing the impact of noise and malfunctioning of components parts. Biologists find such redundancies in nature and engineers build them into systems whose failure may be costly. Managers of firms collect redundant pieces of information so as to reduce the impact noise.

Seventh, why is there a tendency to replace markets with planning during wars? This is because the importance of speed increases during war, and therefore optimal form of economic organization changes. Scitovsky (1951) notes that markets take considerable time to bring about changes in demand and supply. It is no surprise then that during World War II "Everywhere the price mechanism came to be regarded as a method of allocating resources which was too slow and too risky" (Milward 1979, p. 99).

[7]This point is courtesy Peter Boettke.

7 Future Work

Much work remains to be done. *First*, the consequences of markets using prices and firms using communication technology to aggregate information needs to be worked out. This is both a theoretical and modeling challenge.

Second, the architecture that markets and firms use to reduce the impact of noise remains to be studied. A variety of market structures exist in the real world, the rules of interaction in the New York Stock Exchange is very different from those in Marseille's fish market. So are rules of interaction in first and second price auctions. As to how these rules affect the aggregation noise is unclear. The presence of multiple producers of the same product can be thought of as a way markets incorporate redundancy to reduce the impact of noise. Firms create hierarchical structures for the same purpose. Moreover, individuals in markets do not interact through random matching, rather they interact through a network of relationships. Similarly, in firms individuals meet through a web of formal and informal relationships. The impact of interaction topology on aggregate outcome both in firms and markets remains to be studied (Axtell 2001).

Third, a variety of empirical work to illustrate speed and efficiency with which markets and firms operate remain to be done. Case studies of activities going from the market to inside a firm (vertical integration) and vice-versa may be particularly illuminating.

Fourth, in the real world firms and markets interact. Market prices enter firm decision making, and markets themselves are nothing but connections between many firms and consumers. The model should allow individual agents to spontaneously form firms or work through markets, i.e. economic organization ought to be an emergent property of the system. The model then needs to be calibrate to reproduce real world data like firm size distribution.

Fifth, so far it has been assumed that computation is costless.[8] Nothing can be further from the truth. In markets agents use price signals and non-price information to make decisions, after all non-equilibrium prices are not sufficient statistic. This means processing a fair bit of information. In firms, managers have to solve difficult optimization problem and make day to day decisions. The computational complexity of these problem is a matter of economic significance and maybe a determinant of the optimal form of economic organization.

Sixth, how does the number of agents affect the performance of firms and markets in terms of time and efficiency?

---

[8]This point is courtesy an anonymous referee.

Concluding Remarks
Since the 1970s theories of firms based on asymmetric information between agents have displaced the traditional view of the firm as a black-box that transforms inputs into output (Greenwald and Stiglitz 1990). Yet information asymmetries are as much a part of markets as firms (Akerlof 1970), and there are variety of market solutions to problems that arise out of information asymmetry. Therefore, asymmetry of information per se says little about economic organization. The theory propounded in this essay studies the way in which dispersed information is communicated in an economic system. Firms and markets are different means of communication. While firms communicate faster than markets, thereby saving time for other uses, markets are more robust when faced with noise. In a firm information is first conveyed to a centralized authority, who then tells each agent what to do. Markets involve decentralized communication through agent-level interactions. The choice of which mechanism to use depends on the tradeoff between speed and robustness for the economic activity at hand. The existence of variation in the nature of information to be transmitted and the value of speed across economic activities means that the optimal form of economic organization is unlikely to be a corner solution.

References

Akerlof GA (1970) The market for lemons: quality uncertainty and the market mechanism. Q J Econ 84:488–500

Alchian AA, Harold D (1972) Production, information costs, and economic organization. Am Econ Rev 62(5):777–795

Axtell R (2000) Why agents?: on the varied motivations for agent computing in the social sciences. Cent Soc Econ Dyna, Working Paper 17. Brookings Institution

Axtell R (2001) Effects of interaction topology and activation regime in several multi-agent systems. Springer, Berlin

Axtell RL (2003) Economics as distributed computation. Springer, Tokyo, pp 3–23

Bolton P, Joseph F (1990) Decentralization, duplication, and delay. J Polit Econ 98(4):803

Coase RH (1937) The nature of the firm. Economica 4(16):386–405

Coase RH (1988) The nature of the firm: influence. J Law Econ Organ 4(1):33–47

Feldman MS, James GM (1981) Information in organizations as signal and symbol. Admin Sci Q 26:171–186

Greenwald BC, Joseph ES (1990) Asymmetric information and the new theory of the firm: financial constraints and risk behavior. Am Econ Rev 80(2):160–165

Hart O (1989) An economist's perspective on the theory of the firm. Columbia Law Rev 89(7):1757–1774

Hayek FA (1945) The use of knowledge in society. Am Econ Rev XXXV(4):519–530

Hurwicz L (1973) The design of mechanisms for resource allocation. Am Econ Rev 63(2):1–30

Jensen MC, William HM (1976) Theory of the firm: managerial behavior, agency costs and ownership structure. J Financ Econ 3(4):305–360

Marschak T (1959) Centralization and decentralization in economic organizations. Econ J Econ Soc 27:399–430
Milward AS (1979) War, economy, and society, 1939–1945, vol 5. Univ of California Press, California
Myerson RB (2008) Perspectives on mechanism design in economic theory. Am Econ Rev 98:586–603
Polanyi M (1966) The tacit dimension. Univ of Chicago Press, London
Rubinstein A (1998) Modeling bounded rationality, vol 1. MIT Press, Cambridge
Sah R, Joseph ES (1987) The architecture of economic systems: hierarchies and polyarchies. NBER, Working Paper Series No. 1334. http://www.nber.org/papers/w1334.pdf
Scitovsky T (1951) Mobilizing resources for war: the economic alternatives. McGraw-Hill, New York
Simon HA (1951) A formal theory of the employment relationship. Econ J Econ Soc 19:293–305
Simon HA (1962) The architecture of complexity. Proc Am Philos Soc 106(6):467–482
Stiglitz JE (1989) Markets, market failures, and development. Am Econ Rev 79(2):197–203
Von Neumann J (1951) The general and logical theory of automata. Cerebral mechanisms in behavior. Wiley, New York, pp 1–41
Williamson OE (1975) Markets and hierarchies. Free Press, New York, pp 26–30

The "Win-Continue, Lose-Reverse" Rule in Cournot Oligopolies: Robustness of Collusive Outcomes

Segismundo S. Izquierdo and Luis R. Izquierdo

1 Introduction and Motivation

It is generally recognised that the actual decision-making processes followed by real-world firms when they have to set prices or production levels have often little to do with those assumed in the idealized analytical framework of perfect information.[1] In practice, the use of simple revisable strategies, imitation tactics and rules of thumb seems to be a key ingredient of many decision processes. Thus, when analysing a market and its expected behaviour, it seems valuable to go beyond the perfect-information analysis, and also consider other decision procedures that enjoy greater empirical support and which may be deemed plausible for the context at hand. This point is particularly relevant in markets potentially subject to regulation (e.g. oligopolies) and in situations where the perfect-information theoretical analysis of the social interaction reveals the presence of multiple possible equilibria—as is often the case in indefinitely repeated strategic interactions, including oligopolies in particular. Consequently, several different rules for setting prices or production levels in oligopolies have been analyzed. Bigoni and Fort (2013) provide a recent review of the theoretical and experimental literature on learning in oligopolies.

[1]This statement does not necessarily imply that market predictions made using the perfect-information model are irrelevant in real life; the famous "as if" theory of Friedman (Friedman 1953) proves sufficiently accurate and useful in many contexts.

S.S. Izquierdo (✉)
Department of Industrial Organization, Universidad de Valladolid, 47011 Valladolid, Spain
e-mail: segis@eii.uva.es

L.R. Izquierdo
Department of Civil Engineering, Universidad de Burgos, 09001 Burgos, Spain
e-mail: luis@izquierdo.name

F. Amblard et al. (eds.), *Advances in Artificial Economics*, Lecture Notes in Economics and Mathematical Systems 676, DOI 10.1007/978-3-319-09578-3_3

In this paper we analyse Cournot oligopolies in which some firms provide a homogeneous good or service and have to choose their production level q_i. We consider that the market process advances in discrete time steps and at every time step the companies have to simultaneously choose whether to increase or decrease the value of their decision variable (q_i). The decision rule considered here can be simply stated as: repeat your last action (i.e. an increase or a decrease in production) if your profits have grown; otherwise, choose the opposite action. This simple rule has been named "Win-Continue, Lose-Reverse" (WCLR) by Huck et al. (2003),[2] who conducted a thorough study of its convergence properties in symmetric Cournot oligopolies.

The WCLR rule adjusts the level of the decision variable in the direction that is expected to make profits grow, according to the observed effect on profits of the last increment/decrement. Note that this gradual adjustment strategy can be considered a type of reinforcement learning rule: an action (i.e. an increase or decrease in production) is deemed satisfactory—and therefore repeated—if it provides a profit boost, and it is considered unsatisfactory—and therefore avoided—otherwise.

Mathematically, the WCLR strategy presents some similarities with a gradient ascent optimization method. In fact, if the profits of a company were to depend only on its own price or level of production (as in a monopoly with stable demand and costs), this rule would be a gradient ascent method and, under conditions that are well known in the optimization literature (Snyman 2005), it would lead to the vicinity of a local optimum. In a duopoly, however, the profits of a company depend also on its competitor's price or output level, and the application of the WCLR rule by each of the companies independently does not constitute a gradient ascent method for the joint profit of the two companies. Thus, it is interesting to study to which reference point of the strategic game (e.g. collusive outcome, competitive outcome, or one-shot Nash equilibrium) such a simple strategy converges, if it does converge to any at all.

For a Cournot duopoly in which companies vary their production levels q_i by a predefined amount δ (step size), Huck et al. (2003) show that, under rather general conditions, for small values of δ, the quantities q_i converge to a small area around the cooperative (collusive) solution. In this paper, we show that the convergence of the WCLR rule to collusive outcomes is not robust to small independent perturbations in the profit functions of the firms (e.g., small independent variations in the cost functions, or small differences on the price received by each company). The existence of such small independent perturbations tends to push the process towards the Nash equilibrium of the one-shot game.

The structure of the remaining of the paper is very simple: in Sect. 2 we present the results for the Cournot model, and then we end with the conclusions.

[2]The same authors use the name "trial and error" in (Huck et al. 2004), where they also present and discuss this learning rule in a discrete-time setup, though the analysis in that paper is focused on a continuous version of the process.

2 Competition in Quantities: Cournot Model

In this section we analyse a Cournot duopoly in which at every time step t ($t = 0$, $1, \ldots$) each company i ($i = 1, 2$) chooses a production level or quantity $[q_i]_t$. The market price $[p]_t$ is the same for both companies and it depends on the total quantity produced by both firms. The amount $[q_i]_t$ is produced on period t with a cost function $C(q)$. The profit for each company on period t is $[\pi_i]_t = [p]_t\,[q_i]_t - C([q_i]_t)$. Incremental values are naturally defined as $[\Delta\pi_i]_t := [\pi_i]_t - [\pi_i]_{t-1}$, for $t > 0$, and initial values at time step 0 are $[\Delta\pi_i]_0 = 0$, and $[\Delta q_i]_0 = 0$.

Let us also define $[s_i]_t :=$ sign $([\Delta q_i]_t\,[\Delta\pi_i]_t)$. Note that $[s_i]_t$ is equal to $+1$ if the last changes in $[q_i]_t$ and $[\pi_i]_t$ took place in the same direction, and $[s_i]_t$ is equal to -1 if such changes went in opposite directions.

For each company i, the production levels are calculated as $[q_i]_{t+1} = max([q_i]_t + [\Delta q_i]_{t+1}, 0)$, starting with some initial positive production level $[q_i]_0$ at time step 0. The decision rule WCLR used to calculate the production increments $[\Delta q_i]_{t+1}$ is implemented as follows:
WCLR Rule:

- If $t = 0$ or $[s_i]_t = 0$, then $[\Delta q_i]_{t+1}$ takes one random value out of the set $\{-\delta_i, 0, \delta_i\}$, where δ_i is the step size.
- Otherwise, $[\Delta q_i]_{t+1} = \delta_i\,[s_i]_t$.

It is also assumed that the process includes some "noise" such that, with a small probability ε for each company in every period, the company will deviate from the value prescribed above for $[\Delta q_i]_{t+1}$ and will take a random choice out of the set $\{-\delta_i, 0, \delta_i\}$. This "decision noise" can represent occasional mistakes or experimentation.

Huck et al. (2003) prove that, with $\delta_i = \delta$, under rather general conditions, if the step size δ and the noise level ε are sufficiently small (but strictly positive), in the long run the process $[q_1, q_2]_t$ will spend most of the time in a small neighbourhood around the collusive outcome, and their simulations show a quick convergence to that situation. The remaining of this section is devoted to show that this convergence can be very sensitive to small independent perturbations in the profit functions of the firms. The reader can run all the simulations reported here using the online model at http://luis.izqui.org/models/wc-lr-cournot/. The computer model has been implemented in NetLogo (Wilensky 1999).

2.1 *The WCLR Rule in the Cournot Model with Noise*

For illustrative purposes we consider a linear inverse demand function: $p = max(100 - (q_1 + q_2), 0)$ and a quadratic cost function: $C(q) = 10q + 0.1q^2$. In this situation, the collusive value for the production of each company, characterized by the first-order conditions $\dfrac{\partial\,(\pi_1 + \pi_2)}{\partial q_i} = 0$, is $q_i = 21.43$, which corresponds to a price level $p = 57.14$. The Cournot equilibrium, characterized by the equations $\frac{\partial \pi_i}{\partial q_i} = 0$, is $q_i = 28.13$, corresponding to a price level $p = 43.75$.

We also set $\delta_i = 0.1$ and $\varepsilon = 0.01$. Initial levels of production $[q_i]_0$ are set randomly in the range [0, 50], but note that the model is ergodic (since $\varepsilon > 0$); thus, its long-run behaviour does not depend on initial conditions.

Departing from the baseline scenario above, we study the sensitivity of the model to three types of noise:

1. "Decision noise", characterised by the parameter ε, as described above.
2. "Cost noise", characterised by the parameter ε_{cost}, and implemented by altering each firm's base cost according to the following formula:

$$c_i = \left(10q_i + 0.1{q_i}^2\right)\ (1 + \varepsilon_{cost}\mathrm{U}_i\,[-1, 1])$$

   where $\mathrm{U}_i[-1,1]$ denotes a continuous uniform random variable with range $[-1,1]$.
3. "Price noise", characterised by the parameter ε_{price}, and implemented by giving each firm i a price p_i according to the following formula:

$$p_i = p\left(1 + \varepsilon_{price}\mathrm{U}_i\,[-1, 1]\right)$$

   where p is the price that corresponds to the total level of output using the inverse demand function. This modified model represents small differences in the price that each company gets for its products, which can be due to a number of different reasons, such as random deviations in the quality of the products of a company with respect to the average quality, different times of arrival at the market (which would allow for some variability in demand), different intermediaries with variable commissions, existence of local markets (which would allow for some variability in price), etc.

Figure 1 below shows a representative run for each of the three types of noise.[3] It is clear that in the absence of cost noise or price noise, the WCLR rule leads to the collusive outcome, as shown by Huck et al. (2003). In stark contrast, small independent perturbations in the cost function or in the price function seem to destabilise the collusive outcome and push the simulation towards the Cournot equilibrium. The sensitivity of the model to perturbations in price seems to be greater than the sensitivity to perturbations in cost.

To study this effect rigorously, we conducted a computational experiment where we explored different values of ε, ε_{cost}, and ε_{price}. For each value of these variables we conducted 100 simulation runs, and for each of the runs we computed the average price in the simulation (taken over 10^5 time steps, and neglecting the first 10^4 time steps). Figures 2, 3 and 4 below show the results obtained.

---

[3] Note that the simulation runs with "cost noise" or "price noise" have $\varepsilon = 0.01$ too, as prescribed in the baseline scenario.

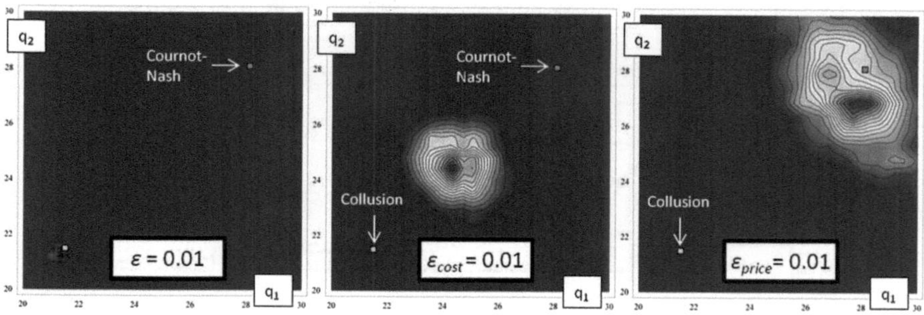

Fig. 1 Density Histograms of the quantities produced by each firm $[q_1, q_2]$ in one representative simulation run of 100,000 time steps. The *left-most histogram* shows a baseline scenario. The histogram in the *centre* corresponds to a simulation run with a 1 % cost noise added to the baseline scenario, whilst the *right-most histogram* shows a simulation run with a 1 % price noise added to the baseline scenario

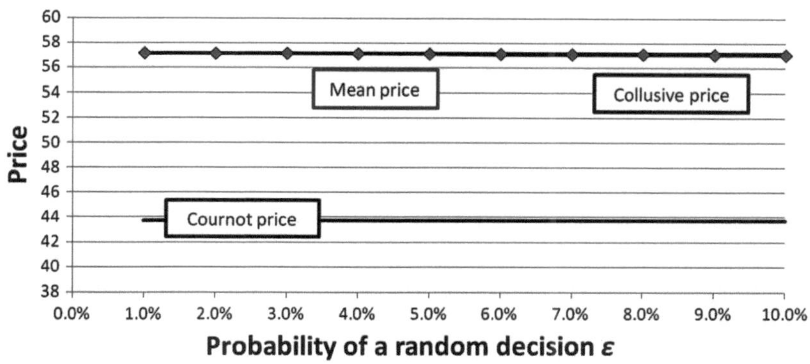

Fig. 2 The *blue diamonds* show, for each value of the probability of a random decision ε, the mean of 100 prices, each of them obtained from one independent simulation run otherwise parameterised as in the baseline case. The price obtained from each simulation run is the average price in that simulation (taken over 10^5 time steps, and neglecting the first 10^4 time steps). The difference between the minimum average price and the maximum average price across simulations was less than 0.1 in all cases

Figure 2 shows that the WCLR rule leads to collusive outcomes even if the probability of a random decision is fairly high. Figure 3, in contrast, shows that small perturbations in the cost functions of the firms destabilise the collusive outcome and push the process towards the Cournot–Nash equilibrium of the one-shot game. In the same spirit, Fig. 4 shows that the sensitivity of the model to small perturbations in prices is even higher, and the collusive outcome is completely destabilised in favour of the Cournot–Nash equilibrium for values of the price noise as low as 1 %.

Why is the WCLR rule so robust to "decision noise", but so sensitive to "cost noise" and "price noise"? Note that the stability of the collusive outcome induced by the WCLR rule relies on coordinated moves. When WCLR firms move in the same direction (either increasing or decreasing production levels), they receive

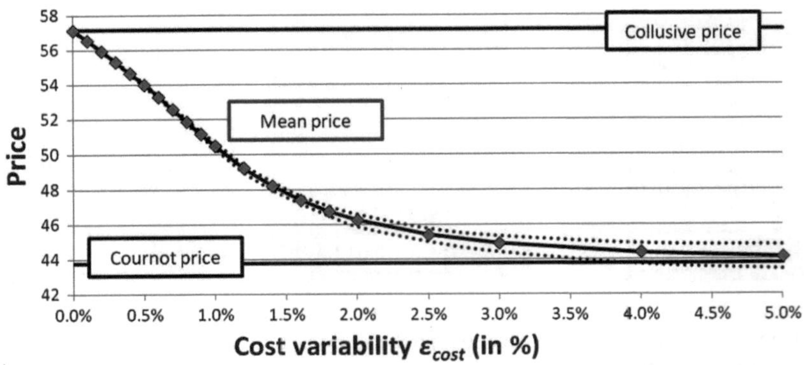

Fig. 3 The *blue diamonds* show, for each value of the cost noise parameter ε_{cost}, the mean of 100 prices obtained from 100 independent simulation runs otherwise parameterised as in the baseline. The price obtained from each simulation run is the average price in that simulation (taken over 10^5 time steps, and neglecting the first 10^4 time steps). The *dashed lines* join the minimum average prices and the maximum average prices across simulations

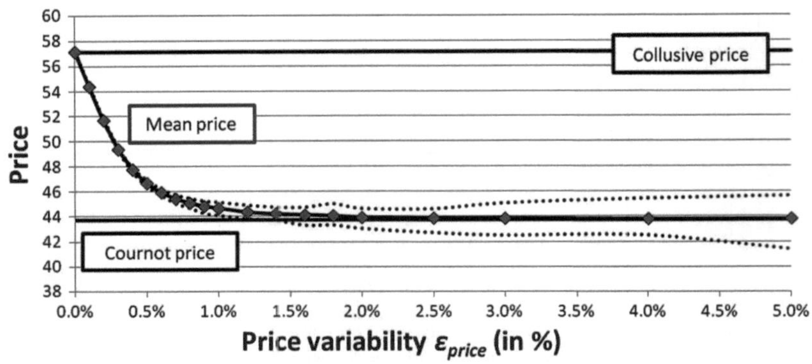

Fig. 4 The *blue diamonds* show, for each value of the price noise parameter ε_{price}, the mean of 100 prices obtained from 100 independent simulation runs otherwise parameterised as in the baseline. The price obtained from each simulation run is the average price in that simulation (taken over 10^5 time steps, and neglecting the first 10^4 time steps). The *dashed lines* join the minimum average prices and the maximum average prices across simulations

signals that make them move towards the collusive outcome and linger around it. Alternatively, an uncoordinated move in the vicinity of the collusive equilibrium (possibly due to a perturbation) will make both firms move towards the Cournot–Nash equilibrium in the following time step—assuming no more deviations from the WCLR rule occur. Note, however, that this move towards Nash is itself coordinated, so at the following time step, both firms will simultaneously decrease production and they will keep doing so until they return to the neighbourhood of the collusive outcome. This explains why the collusive outcome is so robust to “decision noise”.

Decision mistakes have an impact only on the decision at the time step at which they occur, and the process goes back towards collusion automatically in two time-steps.

By contrast, the effects of "cost noise" and "price noise" are more profound, as they do not only affect the decision at the time step they occur, but they also have a direct impact on subsequent decisions. This is because these perturbations effectively change the profit landscape and, in that way, they alter the relation between $[\Delta q_i]_t$ and $[\Delta \pi_i]_t$. This deeper type of perturbation, which transcends the time step at which it occurs, is a greater source of miscoordination and, as explained above, uncoordinated moves push the process towards the Cournot–Nash equilibrium.

One final question remains to be answered: why does price variability have a greater impact than cost variability? The answer relates to the different strength with which these two sources of miscoordination affect the profit landscape. It turns out that, given the parameter values used in the illustrations above, profits for both firms in the region of interest are always positive and quite sizable, i.e. income is significantly greater than cost for both firms. In such a situation, a certain percentage change x in prices (and, therefore, in income) induces a greater change in profit than the same percentage change x in costs. Greater changes in profit mean higher chances of altering the sign of $[\Delta \pi_i]_t := [\pi_i]_t - [\pi_i]_{t-1}$, and hence, greater impact on the dynamics of the model. Thus, under such favourable circumstances, it is natural that price variability constitutes a greater source of miscoordination than cost variability. If income and cost were closer in magnitude, the sensitivity of the model to these two types of noise—"price noise" and "cost noise"—would also be more alike. This point can be checked adding a fixed cost equal to 900, i.e. the new cost function reads $C(q) = 900 + 10q + 0.1q^2$. This change makes income and cost similar in the region of interest. In these conditions, the observed impact of cost variability was similar to that of price variability.

2.2 *Other Noise Distributions*

In this section we show that our results are robust to changes in the noise distribution considered for the price or the cost perturbations. To illustrate this fact, we focus here on a normal distribution with the same mean and standard deviation as the uniform distribution U[−1,1], i.e. the normal distribution N[0, 1/3] with mean 0 and variance 1/3.

First, we show in Fig. 5 below a representative run for each of the three types of noise.[4] Figure 5, which uses the noise distribution N[0, 1/3] for the price and the cost perturbations, is analogous to Fig. 1, which used the noise distribution U[−1,1].

[4]Note that the simulation runs with "cost noise" or "price noise" have $\varepsilon = 0.01$ too, as prescribed in the baseline scenario.

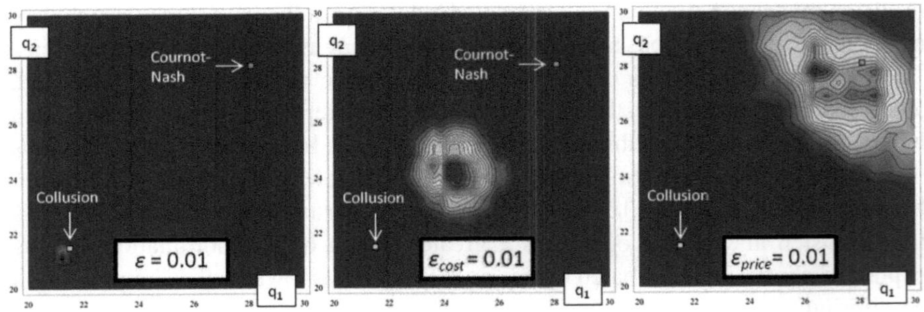

Fig. 5 Density Histograms of the quantities produced by each firm $[q_1, q_2]$ in one representative simulation run of 100,000 time steps. The *left-most histogram* shows a baseline scenario. The histogram in the *centre* corresponds to a simulation run with a 1 % cost noise following a N[0, 1/3] added to the baseline scenario, whilst the *right-most histogram* shows a simulation run with a 1 % price noise following a N[0, 1/3] added to the baseline scenario

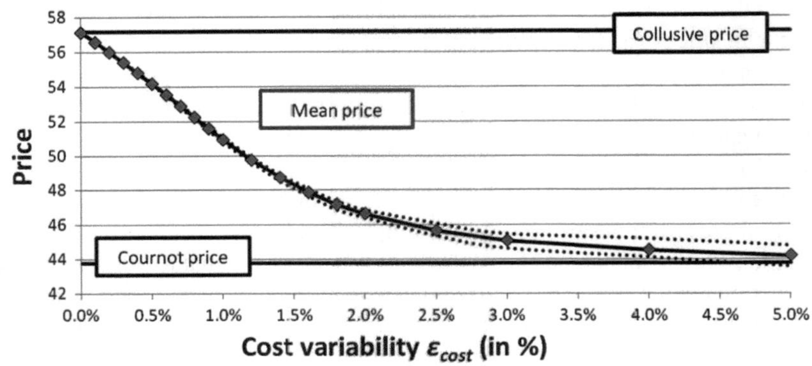

Fig. 6 The *blue diamonds* show, for each value of the cost noise parameter ε_{cost}, the mean of 100 prices obtained from 100 independent simulation runs with noise distribution N[0, 1/3], and otherwise parameterised as in the baseline. The price obtained from each simulation run is the average price in that simulation (taken over 10^5 time steps, and neglecting the first 10^4 time steps). The *dashed lines* join the minimum average prices and the maximum average prices across simulations

To study the robustness to changes in the noise distribution, we conducted a computational experiment where we explored different values of ε_{cost} and ε_{price} using the noise distribution N[0, 1/3], in the same spirit as the experiments shown in Figs. 3 and 4 for noise distribution U[−1,1]. Figure 6 below presents the results obtained for ε_{cost}, which are very similar to those obtained in Fig. 3. The same similarity was obtained for price perturbations (ε_{price}), showing that the sensitivity of the model to cost and price noise does not depend on whether the noise distribution is a uniform distribution or a normal distribution.

2.3 Correlated Perturbations

In this section, we show that the destabilizing factor of the variability in cost or price is not so much the existence of the perturbations, but the fact that they are somewhat independent or uncorrelated between the firms. To illustrate this, here we consider the effect of *correlated* perturbations. Correlations would be observed in the real world if there were variations in costs or in the demand function that affected both companies in a similar way (for instance, seasonal demand variability). To study such situations, we model a price perturbation for each firm which is composed of both a common factor α U[−1,1]—with weight α—and an independent factor $(1-\alpha)$ $U_i[-1,1]$—with weight $(1-\alpha)$—, according to the formula:

$$p_i = p\left(1 + \varepsilon_{price} \mathrm{R}^{\alpha}{}_i\right)$$

where

$$\mathrm{R}^{\alpha}{}_i = \alpha\,\mathrm{U}[-1,1] + (1-\alpha)\;\mathrm{U}_i\,[-1,1].$$

Thus, parameter α is a measure of the correlation between the perturbations of each firm. Extreme value $\alpha = 0$ represents completely uncorrelated perturbations (as analyzed above), and extreme value $\alpha = 1$ represents full correlation (where the perturbations for each firm are exactly the same). Figure 7 below shows that the more correlated perturbations are, the less impact they have on destabilising the collusive outcome. As explained before, perturbations affect the dynamics of the model mainly through the generation of miscoordination between the firms; thus, it is natural that the impact of correlated noise, which does not cause so much miscoordination, is less acute than the effect of uncorrelated perturbations.

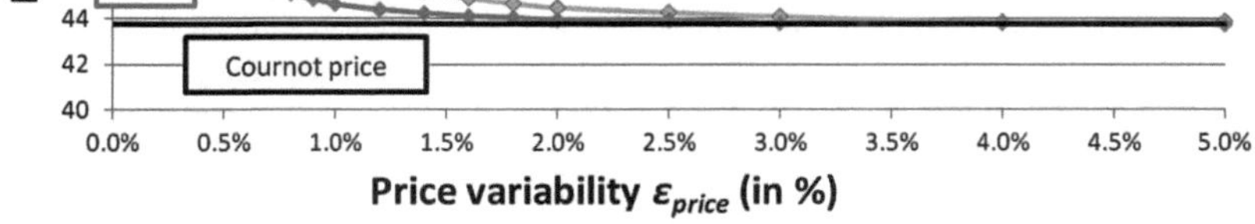

Fig. 7 The *diamonds* show, for each value of the price noise parameter ε_{price} and different values of α, the mean of 100 prices obtained from 100 independent simulation runs otherwise parameterised as in the baseline. The price obtained from each simulation run is the average price in that simulation (taken over 10^5 time steps, and neglecting the first 10^4 time steps)

2.4 More than Two Competing Firms

The simulation results of Huck et al. (2003) in symmetric oligopolies with more than two competing firms (up to ten) and some small decision noise also showed convergence of the WCLR rule to collusive outcomes. We show in Fig. 8 below that, as in the duopoly case, the existence of small independent perturbations in the price that each company obtains also destabilises the collusive outcome and pushes the process towards the Nash equilibrium of the one-shot game.

Uncorrelated perturbations in cost have the same qualitative effect, so they are not shown here.

It should also be noted that, as the number of competing firms increase, the one-shot Cournot–Nash equilibrium gets closer to the outcome predicted under the assumption of perfect competition, so, as the number of firms increase, the WCLR rule with independent cost or price perturbations leads to market prices and production levels which approach those predicted by the perfect competition theory. Figure 9 below shows the effect of uncorrelated 2 % price perturbations in oligopolies with different number of firms. The results also show an increasing difference between the simulated price and the Cournot price as the number of firms in the market increases, which can be due to the decreasing marginal importance of one firm in the market as the number of firms in the market increases.

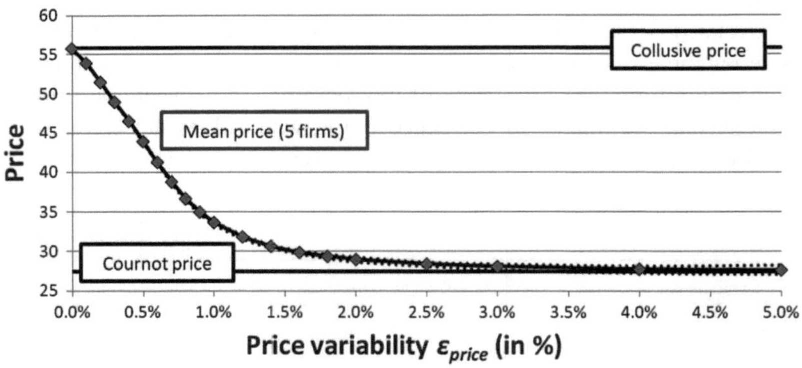

Fig. 8 The *blue diamonds* show, for each value of the price noise parameter ε_{price}, the mean of 100 prices obtained from 100 independent simulation runs otherwise parameterised as in the baseline, in an oligopoly with five competing firms. The price obtained from each simulation run is the average price in that simulation (taken over 10^5 time steps, and neglecting the first 10^4 time steps). The *dashed lines* join the minimum average prices and the maximum average prices across simulations

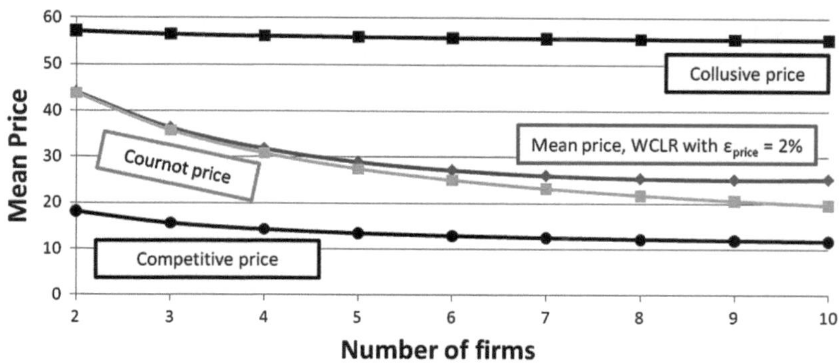

Fig. 9 The *diamonds* show, for a price noise parameter $\varepsilon_{price} = 2\,\%$ and different number of firms, the mean of 100 prices obtained from 100 independent simulation runs otherwise parameterised as in the baseline. The price obtained from each simulation run is the average price in that simulation (taken over 10^5 time steps, and neglecting the first 10^4 time steps)

Conclusions

The results obtained by Huck et al. (2003) indicate that the simple, individual, "sensible" and not forward-looking decision rule WCLR ("Win-Continue, Lose-Reverse") can lead to collusion-like outcomes in Cournot oligopolies, even though each company is independently trying to maximize its own profit, and is acting based only on its own past information. Similar results were obtained by Waltman and Kaymak (2008) considering a more involved learning algorithm (Q-learning). In principle, these results could raise important concerns about the fairness of fining firms in oligopolies for apparently carrying out collusive practices, since one could always allege that observed collusion-like outcomes could just be the unintended result of using this type of independent (and thus legitimate) decision rule.

However, this paper has shown that small independent variations in the cost functions, or small uncorrelated perturbations in the price obtained by each firm, can all destabilize the convergence of the WCLR rule to collusive outcomes, pushing the outcomes towards the Nash solution of the one-shot game. Previous simulation results (Keen and Standish 2006) had also indicated that introducing variability in the step sizes used by each company in each period could also push the process towards the Cournot–Nash solution in markets where firms compete in quantities. Consequently, in markets where there is some independent variability over time in the profit functions of the competing firms (which can be due, for instance, to spatially local effects), our results throw doubts on the validity of arguments that try to justify collusive-like outcomes as the unintended result of this kind of "innocent" decision rules.

Acknowledgments We thank three anonymous reviewers for their useful comments. This work has received financial support from the Spanish Ministry of Science and Innovation (CSD2010-00034, SIMULPAST).

References

Bigoni M, Fort M (2013) Information and learning in oligopoly: an experiment. Games Econ Behav 81(1):192–214

Friedman M (1953) Essays in positive economics. Part I – the methodology of positive economics. University of Chicago Press, Chicago, pp 3–43

Huck S, Normann HT, Oechssler J (2003) Zero-knowledge cooperation in dilemma games. J Theor Biol 220(1):47–54

Huck S, Normann HT, Oechssler J (2004) Through trial and error to collusion. Int Econ Rev 45(1):205–224

Keen S, Standish R (2006) Profit maximization, industry structure, and competition: a critique of neoclassical theory. Physica A 370(1):81–85

Snyman JA (2005) Practical mathematical optimization: an introduction to basic optimization theory and classical and new gradient-based algorithms. Springer, Heidelberg

Waltman L, Kaymak U (2008) Q-learning agents in a Cournot oligopoly model. J Econ Dyn Control 32(10):3275–3293

Wilensky U (1999) NetLogo. Center for Connected Learning and Computer-Based Modeling, Northwestern University, Evanston. http://ccl.northwestern.edu/netlogo/

Organizational Change for Its Own Sake?

Results of an Agent-Based Simulation

Friederike Wall

1 Introduction

The dissemination of organizational design elements changes over time, and the popularity of some management techniques seems to follow a cyclic pattern. This gives reason to terms like "management fashions" or "management fads" (Abrahamson 1991, 1996; Kieser 1997). In the last decades, for example, there were several oscillating movements, e.g., towards integration ("mergers and acquisitions") as well as disintegration of firms ("demergers"), the rise and fall of quality circles and business process reengineering (Dale et al. 2001); the prevalence of employee-stock-ownership programs increased and decreased regularly (Abrahamson 1996), and we saw several swings between centralization and decentralization (Mintzberg 1979).

It has been discussed whether "fashion-like" organizational changes are driven by an interplay between performance gaps due to changes in the economic or technological environment, by changes of "aesthetic" preferences or by socio-psychological factors affecting managers (Abrahamson 1996). A related question is whether the "cyclic" prevalence of certain organizational design elements is beneficial with respect to organizational performance. This issue is not directly addressed in the "management fashion"-literature; however, there is a large body of research on the fit between environmental dynamics, complexity and organizational structures (e.g., Levinthal 1997; Rivkin 2001; Siggelkow and Levinthal 2003; Siggelkow and Rivkin 2005). According to this literature, when adapting to their environment, organizations are subject to a tension between exploration and

F. Wall (✉)
Alpen-Adria-Universität Klagenfurt, Klagenfurt, Austria
e-mail: friederike.wall@uni-klu.ac.at

F. Amblard et al. (eds.), *Advances in Artificial Economics*, Lecture Notes in Economics and Mathematical Systems 676, DOI 10.1007/978-3-319-09578-3_4

exploitation, i.e., between discovering new solutions and increasing the yield of well-known solutions—or in other words: between search and stability.

Against this background, in this paper we investigate, whether, or not, the sheer occurrence of regular changes in organizational design could induce improvements in organizational performance—and, by that, could explain cyclic ups and downs of organizational design patterns.

For this, we employ an agent-based simulation model based on the idea of NK fitness landscapes (Kauffman 1993; Kauffman and Levin 1987). In particular, we observe artificial organizations in their effort to increase long-term organizational performance. Our organizations simultaneously use two means—first, in short-term they search stepwise for better solutions to a given decision-problem, and second, in mid-term they can change their organizational structure.

An agent-based simulation approach appears an appropriate research method for our research question since it allows studying alternative modes of organizational change in a procedural perspective and to consider the collaboration of parties, like departments, central unit, managers, within an organization (Chang and Harrington 2006). Obviously, it would be rather difficult to obtain empirical data in the required longitudinal perspective and to control for the components of organizational change; moreover, the complexity of components to be considered in a processual view would lead to rather intractable dimensions in closed-form modeling (Davis et al. 2007; Harrison et al. 2007).

The remainder of the paper is organized as follows: Subsequently, Sect. 2 outlines major aspects of the simulation model and, in Sect. 3 we shortly present some of the results.

2 Outline of the Simulation Model

Our simulation is based on the idea of NK fitness landscapes, and we observe organizations searching for superior levels of organizational performance within adaptive walks on these landscapes. In doing so, we rely on a simulation approach (Davis et al. 2007) which has been widely used in managerial science (for an overview see, e.g., Ganco and Hoetker 2009; Sorenson 2002). However, the distinctive feature of our model is that we employ—in addition to a short-term search for superior performance at a given organizational setting—different forms of mid-term dynamics in which the organizational setting is modified.

2.1 A Model of Interrelated Decisions and Delegation

In each time step t of the observation period, our artificial organizations face an N-dimensional binary decision problem, i.e., they have to make decisions $d_{it} \in \{0, 1\}$, $(i = 1, \ldots N)$. Hence, the search space at each time step consists of

2^N different binary vectors $\mathbf{d}_t \equiv (d_{1t}, \ldots, d_{Nt})$ possible. Each of the two states $d_{it} \in \{0; 1\}$ makes a certain contribution C_{it} to overall performance $V(\mathbf{d}_t)$ of the organization. C_{it} is randomly drawn from a uniform distribution with $0 \leq C_{it} \leq 1$.

The NK framework allows for representing interactions among decisions with level K. K reflects the number of other choices d_{jt}, $j \neq i$ which also affect the performance contribution C_{it} of decision d_{it}. K can take values from 0 (no interactions) to $N-1$ (maximum interactions). With this, performance contribution C_{it} might not only depend on the single decision d_{it} but also on K other decisions d_{jt} where $j \in \{1, \ldots N\}$, and $j \neq i$.

$$C_{it} = f_i\left(d_{it}, d_{jt, j\in\{1,\ldots N\}, j\neq i}\right) \tag{1}$$

The overall organizational performance V_t achieved in period t results as normalized sum of performance contributions C_{it} from

$$V_t = V\left(\mathbf{d}_t\right) = \frac{1}{N}\sum_{i=1}^{N} C_{it} \tag{2}$$

Our organizations segment their N-dimensional decision problem into M disjoint partial problems and delegate each of these sub-problems to one department subscripted by r, $r = 1, \ldots, M$ correspondingly. Each department has primary control of its "own" subset of the N decisions. Hence, from the perspective of department r the organizational decision problem is partitioned into a partial decision vector $\mathbf{d}_t^r$ for those decisions which are in the "own" responsibility and into $\mathbf{d}_t^{rRES}$ for the residual decisions that the other departments $q \neq r$ are in charge of. However, in case of cross-departmental interactions, choices of department r might affect the contributions of the other departments' choices and vice versa.

The elements of the model described so far, capture the size N of the decision problem $\mathbf{d}$ an organization faces, the level K of interactions among the single decisions, the segmentation of $\mathbf{d}$ into sub-problems and the sub-problems' delegation to r departments and, finally, how the organizational performance V results from a certain configuration $\mathbf{d}$. In our simulations, these features remain stable over the entire observation period T. Next we turn to the short-term and the mid-term dynamics of the organizations mapped in our simulation model.

2.2 Short-Term Dynamics

In each time step t, a department head seeks to identify the best configuration for the "own" subset of choices $\mathbf{d}_t^r$ assuming that the other departments do not alter their prior subsets of decisions. In particular, a department head randomly discovers two alternative partial configurations of those binary decisions he/she is in charge of: an alternative configuration that differs in one decision (*a1*) and another alternative (*a2*) where two bits are flipped compared to the current configuration. Hence, together with the status quo $\mathbf{d}_{t-1}^{r*}$ and the two alternatives $\mathbf{d}_t^{r,a1}$ and $\mathbf{d}_t^{r,a2}$ head of department r has three options to choose from, and (according to economic literature) favors that option which he/she perceives to promise the highest value base for compensation. As is common for adaptive walks we use a hill-climbing algorithm for capturing this in the simulation model.

The value base for compensation is defined in the incentive scheme. The incentive scheme shows a linear additive relation for two components: first, the normalized sum of performance contributions resulting from the N^r "own" decisions Eq. (3), and second, the performance achieved from the rest of the organization Eq. (4):

$$B_t^{rOWN}\left(\mathbf{d}_t^r\right) = \frac{1}{N} \cdot \sum_{i=1+p}^{N^r} C_{it}$$
$$\text{with} \quad p = \sum_{s=1}^{r-1} N^s \;\text{ for }\; r > 1 \;\text{ and }\; p = 0 \;\text{ for }\; r = 1 \tag{3}$$

$$B_t^{rRES} = \sum_{\substack{q=1 \\ q \neq r}}^{M} B_t^{qOWN} \tag{4}$$

The overall compensation results as given in Eq. (5) where parameter α^r determines to which extent the rest of the organization's performance affects head of department r's value base for compensation:

$$B_t^r\left(\mathbf{d}_t\right) = B_t^{rOWN}\left(\mathbf{d}_t^r\right) + \alpha^r \cdot B_t^{rRES}\left(\mathbf{d}_t\right) \tag{5}$$

Hence, in case of cross-departmental interactions, unit r is also able to affect its "residual" performance contribution. Thus, it depends on the value of α^r whether, for example, only the department's own performance ($\alpha^r = 0$) or firm performance ($\alpha^r = 1$) is rewarded and, by that, which of the options $\mathbf{d}_{t-1}^{r*}$, $\mathbf{d}_t^{r,a1}$ and $\mathbf{d}_t^{r,a2}$ appears favorable in each period t.

However, in our model decision-makers might suffer from certain informational imperfections. In particular, we assume that departments decide on basis of the *perceived* value base for compensation rather than the *actual*. Therefore, we "distort" the actual performance contributions according to the expertise of each single department. A common idea of many organizational theories is that decision-makers in organizations dispose of information with different levels of imperfections (e.g., Ginzberg 1980; Galbraith 1974). For example, departmental decision-makers are assumed to have relatively precise information about their own area of competence, but limited cross-departmental knowledge whereas the main office might have rather coarse-grained, but organization-wide information.

In particular, the perceived value base for compensation is computed as sum of the actual own performance and actual residual performance, respectively each distorted with an error term as given in Eqs. (6) and (7)

$$\tilde{B}_t^r(\mathbf{d}_t) = \tilde{B}_t^{rOWN}(\mathbf{d}_t^r) + \alpha^r \cdot \tilde{B}_t^{rRES}(\mathbf{d}_t) \tag{6}$$

where

$$\begin{aligned} \tilde{B}_t^{rOWN}(\mathbf{d}_t^r) &= B_t^{rOWN}(\mathbf{d}_t^r) + e^{rOWN}(\mathbf{d}_t^r) \\ \tilde{B}_t^{rRES}(\mathbf{d}_t) &= B_t^{rRES}(\mathbf{d}_t) + e^{rRES}(\mathbf{d}_t) \end{aligned} \tag{7}$$

At least with respect to accounting systems (Labro and Vanhoucke 2007), it is reasonable to assume that high (low) true values of performance come along with high (low) distortions. Hence, we reflect distortions as relative errors imputed to the true performance (for other functions Levitan and Kauffman 1995), and, for simplicity, the error terms follow a Gaussian distribution $N(0;\sigma)$ with expected value 0 and standard deviation σ where we differentiate the standard deviation according to specialization of departments. In particular, the standard deviations σ^{rOWN} and σ^{rRES}, for the sake of simplicity, are assumed to be stable in time and the same for all departments r. Errors are assumed to be independent from each other.

Hence, apart from the incentives given (namely parameter α^r), also the managers' individual views according to their expertise affect which of the options $\mathbf{d}_{t-1}^{r*}$, $\mathbf{d}_t^{r,a1}$ and $\mathbf{d}_t^{r,a2}$ is favored in period t.

2.3 Mid-term Dynamics

In the very core of our modelling effort is that our organizations can change their organizational structure from time to time. In particular, two "modes" of mid-term dynamics are compared against each other—and against keeping the structure stable for the entire observation period T:

- The organizational structure is changed *periodically* after T^* periods regardless of, for example, the performance level achieved.
- The organizational structure is changed *value driven*, i.e., in every T^*-th time step the performance change $\Delta V = V_t - V_{t-T^*}$ is assessed and, if ΔV is below a certain threshold v, then the organization is changed.

In particular, changes are put forward along three dimensions of organizational design:

- The *knowledge base* can contain rather expert-like knowledge (being highly precise with respect to departmental performance, but rather coarse-grained for the performance achieved in the rest of the organization) or in a generalist-like structure with medium-precise information on the entire organization (Wall

2011). The level of expertise the department heads have is defined by the error terms $e_t^{rOWN}(\mathbf{d}_t^r)$ and $e_t^{rRES}(\mathbf{d}_t)$ in Eq. (7), and, in particular, by the related standard deviations σ^{rOWN} and σ^{rRES} as described in Sect. 2.2. The idea of how the change in the knowledge base from a specialist- to a generalist-like organization could happen is that the departments could be re-arranged in their composition by mutual transfer of personnel.

- The *reward structure* can be switched between rewarding firm performance V_t, departmental performance only or rewarding departmental performance plus a certain ratio of performance achieved by the rest of the organization. In concrete parameter α^r in Eqs. (5) and (6), respectively, is altered.
- The modes of *coordination* among departments can be changed between three modes (for these and other coordination modes see Siggelkow and Rivkin 2005; Dosi et al. 2003):
    - In a fairly decentralized mode each department decides on the "own" partial decisions $\mathbf{d}_t^r$ autonomously, and the overall configuration $\mathbf{d}$ results as a combination of these departmental choices without any central intervention).
    - As a type of horizontal coordination our departments inform each other about their preferences, and the departments are allowed to veto laterally against each other mutually.
    - In a rather central mode of coordination each department transfers a list with the two most preferred options from $\mathbf{d}_{t-1}^{r*}$, $\mathbf{d}_t^{r,a1}$ and $\mathbf{d}_t^{r,a2}$ to the main office. The main office chooses that combination of the r lists of preferences received that promises the highest overall performance. However, the main office also might suffer from error in ex ante-evaluation of options (also following a Gaussian distribution with mean 0, see Sect. 2.2).

In the periodical as well as the value-driven mode, changes in two *randomly* selected dimensions of these three dimensions (i.e., knowledge base, incentive scheme, coordination mode) can take place. Moreover, the alternative organizational design option within a dimension is also chosen randomly. Additionally, it is worth mentioning, that the initial organizational features according to these dimensions are determined randomly, too.

3 Simulation Experiments and Parameter Settings

For simulating an adaptive walk, after a "true" fitness landscape is generated, the organizations are placed randomly in the fitness landscape and observed while searching for higher levels of overall performance under the regime of the mode of change (periodical, value-driven change, no change) as introduced in Sect. 2.3. Table 1 summarizes the parameter settings.

The simulation experiments are carried for two interaction structures of decisions which, in a way, represent two extremes (for these and other structures see Rivkin and Siggelkow 2007): in the *self-contained* structure intra-departmental interactions

Table 1 Parameter settings

Parameter	Values/types
Observation period	$T = 200$
Number of decisions	$N = 10$
Interaction structures	Self-contained ($K = 4$) Fullinterdependent ($K = 9$)
Number of departments	$M = 2$ with department 1 in charge of partial vector $\mathbf{d}^1 = (d_1, \ldots d_5)$ and department 2 in charge of partial vector $\mathbf{d}^2 = (d_6, \ldots d_{10})$
Incentive structure (i.e., ratio at which residual performance is rewarded	Three levels: $\alpha^r \in \{0; 0.5; 1\}$
Level of errors in terms of standard deviations σ^{rOWN}, σ^{rRES} and σ^{rMAIN} (each with mean zero)	Generalist (0.1; 0.15; 0.125) Specialist (0.05; 0,2; 0.125) Perfect (0; 0; 0)
Coordination mode	Decentralized mode, lateral veto mode, proposal mode
Change mode	Periodically: $T^* = 5$ Value-driven: $v = 0.01$; $T^* = 5$

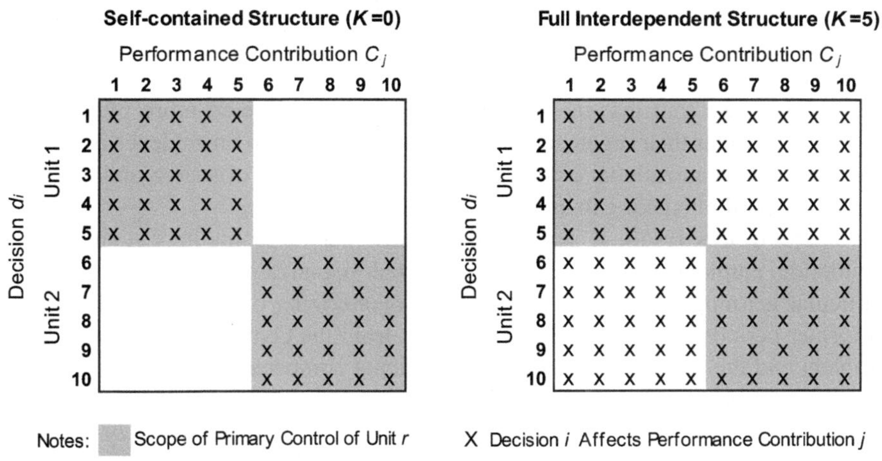

Fig. 1 Interaction structures in the simulation experiments

among decisions are maximal intense while no cross-unit interdependencies exist. In contrast, in the *full interdependent* case all decisions affect the performance contributions of all other decisions, i.e., the intensity of interactions and the coordination need is raised to maximum (Fig. 1).

4 Results and Discussion

Table 2 reports the condensed results of our simulations. In particular, the table displays the performance achieved in the end of the observation period, i.e. $V_{t=200}$, which can serve as an indicator for the effectiveness of the search process as well as the frequency of how often the global maximum in the performance landscape is achieved in the last period observed. The average performance $\overline{V}_{\{0;200\}}$ over the observation time, i.e., achieved by average in each of the 200 periods, might be regarded as a condensed speed measure. Figure 2 displays the adaptive walks under the three change modes under investigation—averaged over the 5,000 adaptive walks simulated for each change mode and each interaction structure.

We discuss the results in three steps. In particular, we compare (1) the two settings including organizational change against the "no change" setting, (2) the periodical against the value-driven change mode and (3) results for the two interaction structures against each other.

Effects of organizational change. For both interaction structures the search processes employing organizational changes are more effective than those in which the organizational structure is kept stable. In particular, the final performance and the average performance in the settings with organizational changes goes beyond the levels achieved in the "no change" settings (with non-overlapping confidence intervals for $V_{t=200}$ for both interaction structures and in all but one cases for $\overline{V}_{\{0;200\}}$).

Moreover, we find that the global maximum is discovered more often if the organizations alter their organizational structure from time to time. Hence, we argue that frequent organizational changes increase the diversity of search and, by that, reduce the peril of sticking to an inferior local maximum.

Regarding this result, it should be stressed that these findings stem from a simulation employing only randomly selected organizational changes, i.e., no particular learning or imitation strategies or memorizing of successful alterations in former periods take place. Moreover, apart from the complexity of the decision problem and the scope of the subunit's decisional competencies the other features of the organizational structures were initiated by random choice.

Hence, the results let us hypothesize, that *the sheer occurrence of frequent organizational change enhances organizational performance.*

Effects of the change mode. In comparing the value-driven against a purely time-driven strategy of change for a given interaction structure, we find that the final performances achieved under the regime of the two change modes are at nearly the same levels. The similar also applies to the frequency of how often the global maximum is found in the last period observed.

However, differences between the change modes show up with respect to the frequency of changes (right most column in Table 2). In the value-driven mode the number of changes is around three-quarters of those in the periodical mode. In a more detailed analysis of the simulation results, we found that in the early stages

Table 2 Condensed results

Scenario-name	Final performance $V_{t=200}$	CI* of final performance	Average performance $\overline{V}_{\{0;200\}}$	CI* of average performance	Frequency of_global maximum in $t = 200$ (%)	Frequency of organizational changes
Self contained structure						
No change	0.9451	0.0028	0.9385	0.0027	25.2	0
Value-driven change	0.9657	0.0022	0.9547	0.0012	38.5	31.55
Periodical change	0.9672	0.0021	0.9557	0.0012	38.8	40
Full interdependent structure						
No change	0.875	0.0034	0.8646	0.0031	4.5	0
Value-driven change	0.883	0.0043	0.8675	0.0023	8.3	30.91
Periodical change	0.8891	0.004	0.8735	0.0022	8.6	40

Notes: *Confidence intervals at a confidence level of 0.001. Each row represents results of 5,000 adaptive walks: 1,000 distinct fitness landscapes with 5 adaptive walks on each; for "Periodical change" $T^* = 5$; for "Value driven change" $T^* = 5$ and $v = 0.01$

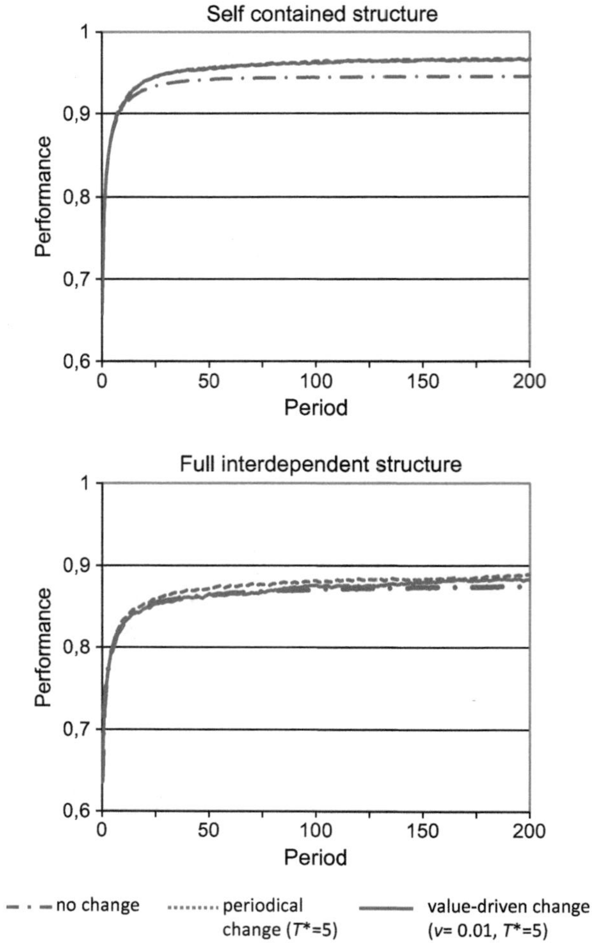

Fig. 2 Performance enhancements with different modes of organizational change

Notes: Each line represents the averaged performance of 5,000 adaptive walks: 1,000 distinct performance landscapes with five adaptive walks on each.

of the search processes (when performance enhancements are particularly high and beyond the threshold v) changes occur rather seldom in the value-driven mode.

Reasonably, organizational changes are not costless (which is not reflected in our model). Hence, the number of changes might be rather relevant with respect to the net-benefit of organizational change. In particular, results suggest that a value-driven change mode is more preferable if the organization is in a stage of dynamic performance enhancements like in the early periods of our observation period.

In sum, this lets us hypothesize that *the value-driven mode is more efficient than the periodical change mode*.

Effects of complexity. The results indicate that frequent organizational change is beneficial in both interaction structures but that it is more beneficial in case of decomposable, i.e. self-contained, structures. In general, changing the organization reasonably re-directs the search and, by that, reduces the threat of sticking to a local maximum and enhances to temporarily break through myopic departmental egoism. In this sense our results are in line with findings of Siggelkow and Levinthal (2003).

However, the question remains why the beneficial effect apparently is higher in case of the self-contained structures which to a certain extent runs contrary to intuition since the problem of sticking to local peaks is more relevant in highly rugged landscapes, i.e. complex environments (Rivkin and Siggelkow 2007). We argue that according to prior research (Wall 2010) self-contained interactions appear to be rather sensitive to coordination mechanisms which do not reflect the abundance of cross-unit interaction (i.e., which seek to coordinate where no coordination need exists), and, in particular, if combined with imperfect knowledge bases of decision makers.

Conclusion

Results suggest that organizational change, for example due to management fashions, by itself has performance enhancing effects—even if changes in detail are randomly set-up. On the one hand this could explain the occurrence of management fashions and frequent organizational changes.

On the other hand, in a way, the results might also be somewhat provoking, for example with respect to organizational learning, since the performance enhancing effects were achieved with undirected changes in terms of random choices of organizational alternatives—suggesting that change per se is of value.

Our analysis is subject to some limitations which should be overcome in further research efforts. As such further studies should investigate more into detail the role of the change parameters like, for example, value and time thresholds triggering changes or the dimensionality of change—which all were fixed in the simulation presented in this paper. Moreover, it is to be mentioned that the two change modes introduced here are rather simplistic and should be related to the broad literature on organizational learning.

References

Abrahamson E (1991) Managerial fads and fashions: the diffusion and refection of innovations. Acad Manag Rev 16(3):586–612

Abrahamson E (1996) Management fashion. Acad Manag Rev 21(1):254–285

Chang M-H, Harrington JE (2006) Agent-based models of organizations. In: Tesfatsion L, Judd KL (eds) Handbook of computational economics: agent-based computational economics, vol 2. Elsevier, Amsterdam, pp 1273–1337

Dale BG, Elkjaer MBF, van der Wiele A, Williams ART (2001) Fad, fashion and fit: an examination of quality circles, business process re-engineering and statistical process control. Int J Prod Econ 73(2):137–152
Davis JP, Eisenhardt KM, Bingham CB (2007) Developing theory through simulation methods. Acad Manag Rev 32(2):480–499
Dosi G, Levinthal D, Marengo L (2003) Bridging contested terrain: linking incentive-based and learning perspectives on organizational evolution. Ind Corp Chang 12(2):413–436
Galbraith JR (1974) Organization design: an information processing view. Interfaces 4(3):28–36
Ganco M, Hoetker G (2009) NK modeling methodology in the strategy literature: bounded search on a rugged landscape. In: Bergh DD, Ketchen DJ (eds) Research methodology in strategy and management. Emerald, Bingley, pp 237–268
Ginzberg MJ (1980) An organizational contigencies view of accounting and information systems implementation. Acc Organ Soc 5:369–382
Harrison JR, Zhiang LIN, Carroll GR, Carley KM (2007) Simulation modeling in organizational and management research. Acad Manag Rev 32(4):1229–1245
Kauffman SA (1993) The origins of order: self-organization and selection in evolution. Oxford University Press, Oxford
Kauffman SA, Levin S (1987) Towards a general theory of adaptive walks on rugged landscapes. J Theor Biol 128(1):11–45
Kieser A (1997) Rhetoric and myth in management fashion. Organization 4(1):49–74
Labro E, Vanhoucke M (2007) A simulation analysis of interactions among errors in costing systems. Account Rev 82(4):939–962
Levinthal DA (1997) Adaptation on rugged landscapes. Manag Sci 43(7):934–950
Levitan B, Kauffman SA (1995) Adaptive walks with noisy fitness measurements. Mol Divers 1(1):53–68
Mintzberg H (1979) The structuring of organizations. Prentice Hall, Englewood Cliffs
Rivkin JW (2001) Reproducing knowledge: replication without imitation at moderate complexity. Organ Sci 12(3):274–293
Rivkin JW, Siggelkow N (2007) Patterned interactions in complex systems: implications for exploration. Manag Sci 53:1068–1085
Siggelkow N, Levinthal DA (2003) Temporarily divide to conquer: centralized, decentralized, and reintegrated organizational approaches to exploration and adaptation. Organ Sci 14(6):650–669
Siggelkow N, Rivkin JW (2005) Speed and search: designing organizations for turbulence and complexity. Organ Sci 16(2):101–122
Sorenson O (2002) Interorganizational complexity and computation. In: Baum JAC (ed) Companion to organizations. Blackwell, Oxford, pp 664–685
Wall F (2010) The (beneficial) role of informational imperfections in enhancing organizational performance. In: LiCalzi M, Milone L, Pellizzari P (eds) Progress in artificial economics: computational and agent-based models. Springer, Berlin, pp 101–122
Wall F (2011) Diversity of the knowledge base in organizations: results of an agent-based simulation. In: Demezeau Y, Pechoucek M, Cochado JM, Pérez JB (eds) Advances on practical applications of agents and multiagent systems, vol 88, Advances in intelligent and soft computing. Springer, Berlin, pp 13–20

Best Practices in Programming Agent-Based Models in Economics and Finance

A. Vermeir and H. Bersini

1 Introduction

The agent-based models (ABM) is a recent class of computational tools, simulating the interactions of autonomous intelligent agents in order to analyze the non-trivial outcome of such system as a whole. It easily find applications in diversified or even multi-disciplinary fields: biologic systems, social sciences, finance and economics, etc. At the cost of significant needs in processing power, they offer flexibility and robustness by remaining consistent inducing heterogeneity: varying parameters, hypothesis, or agents behaviors.

However, designing such model requires preparation and methodology:

- Given their flexibility, ABMs are meant to evolve, prone to multiple successive setups. Designers should pay attention to the robustness of their models.
- Also, given the decentralized and heterogeneous nature of the model dynamics, simplicity and clarity should be maintained when possible.

Since the early 1990s, several well-structured ABM framework emerged (Collier 2003), easing the development burden of researchers. However, specific needs might require original designs. Across our recent collaborations, we observed that researchers tend to build their model step-by-step, not conceptualizing beforehand, resulting in a rigid and complex patchwork of interacting entities difficult to understand, to maintain and to modify. In this paper, we will strive to provide recommendations regarding ABMs implementations in economics and finance.

What follows is divided in three part. First, we highlight the synergy existing between ABMs, Object-Oriented programming and Unified Modeling Language,

A. Vermeir (✉) • H. Bersini
Université Libre de Bruxelles, Avenue Franklin Roosevelt 50, 1050 Brussels, Belgium
e-mail: avermeir@ulb.ac.be; aurelien.vermeir@ulb.ac.be; Hugues.Bersini@ulb.ac.be

F. Amblard et al. (eds.), *Advances in Artificial Economics*, Lecture Notes in Economics and Mathematical Systems 676, DOI 10.1007/978-3-319-09578-3_5

then we formulate recommendations to help researchers in conceptualizing the optimal structure regarding their needs. Finally, we illustrate the planning of a typical implementation based on a concrete case.

2 Two Useful Paradigm: Object-Oriented and Unified Modeling Language

As opposed to bulk sequential instructions processing, the Object-Oriented (OO) (Bersini 2013) paradigm regroups the data and capabilities of existing implicit structures in implementations via entities called "objects". Every object created from the same "class" shares the same capabilities, while possessing its own parameters value. Heterogeneous agents fit perfectly to objects description. The translation of an envisioned ABM in OO is therefore often seamless.

Aside from the ease of implementation of ABMs, the benefits of OO programming are multiple: ease to maintain, to develop, to communicate, to split into team work packages, to modularize… at the cost of being slightly more demanding in processing power but this trade-off falls short in significance compared to the optimization routines often performed at each timestep when agents have to make decisions.

Also, programming languages offer OO specific "services" such as automatic deletion of unused objects, polymorphism, etc., for a more flexible and understandable implementation.

The Unified Modeling Language (UML) (Charroux et al. 2013) is a set of high-level diagrams providing a synthesized view of object interactions. In is therefore the tool of choice when communicating about ABMs (Bersini 2012). In our paper, we make extensive use of simplified class diagrams.

In those diagrams, while rectangles are the objects classes, the entities enriched with capabilities. Illustrated in Fig. 1, the arrows symbolize interactions. Associations are temporary interactions. Dependency represents a link between objects having recurrent interactions. The composition ties objects with their container, should the container disappear, the objects would too. Finally, inheritance provides the common capabilities of the parent object to the children.

The novelty, originality and complexity of ABMs benefit from communication paradigms such as UML. Also, diagrams serve as planning tools to structure the implementation once the model has been settled.

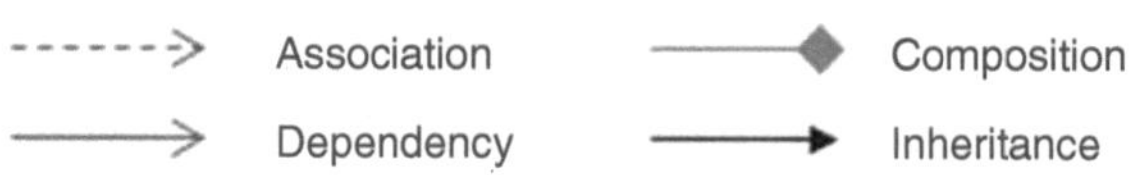

Fig. 1 Illustration of UML relations

3 An Implementation Methodology

3.1 Benefits from Planning

Nowadays, ABM implementations are made using ad hoc tools or generic programming languages. In the first case, researcher benefit from a structured environment and focus on the code. In social science, from the oldest Sugarscape (Epstein and Axtell 1996) to the more recent Repast (Collier 2003), a wide range of frameworks exist.

Our paper focuses on the second case where, depending on needs, complex implementations are realized and where methodology plays a critical role in the evolution of the model. We observed that in many situations, researchers implement the model dynamics step-by-step, resulting in monolithic programs that are a burden to maintain, develop and communicate. When a programmer undertakes an implementation job, he often envisions several coding structures that achieve the desired outcome. During a proper planning phase, he should distinguish a solution from another on the basis of its simplicity, its robustness and its clarity. In Agent-Based Modeling, simplicity and robustness are often a trade-off that depends on the model scope.

- For a pedagogic model, simplicity comes first.
- For a complex research tool, robustness will increase the model modularity and ease its development.
- Finally, the clarity is improved by having an implementation structure closely related to the model.

On the basis on implementations and refactoring of existing ABM models (Bluhm et al. 2012; Georg 2013; Iori et al. 2008), we formulate some advices on the structure and the rationale to prioritise.

3.2 Implementation Recommendations

3.2.1 Focus on Business Logic, Separate it from Technical Parts

ABMs in finance and economics strive to mimic the real world. The OO design of the code should closely follow the same structure. Then, technical parts of the code, such as the data collection, database queries, random number generators, etc. should be in separate classes. Then, the other non-technical classes ("business logic classes") should be structured in a realistic and understandable way:

- Modeling real life business objects (accounts, loans, inventories, etc.)
- Increasing the amount of classes, to ease the code burden and improve clarity, without hampering simplicity

3.2.2 Highlight of Design Patterns

Design patterns are standardized implementations of specific needs. In finance and economics ABM, strategy patterns and observer patterns are very common.

Observer Pattern (Fig. 2) When the model includes some form of timestep involving a schedule of agents' actions, or simply a periodical single trigger for each agent, the observer pattern should be implemented. Here, we suppose a simulation creating agents and adjuvants (non-agent actors). The simulation must provide their references to the scheduler object. Then, the scheduler, implementing the interface Observable, should have a "notify" method that triggers the "update" method for each registered object. Having those objects implementing the Observer interface guarantees the existence of such update method.

Strategy Pattern (Fig. 3) The strategy pattern separates the behavior of an agent from the rest of its dynamics in two classes:

- The agent class should include the capabilities of the agent
- The strategy classes should contain the sequences of actions ("strategies") of the agent methods

Implementing it is a plus in most cases and a must if each agent has multiple strategies or if strategies are varying among agents. Typically, the simulation creates agents and strategies objects. During the simulation, each agent is given a strategy and executes it. In each strategy object, a list of method calls trigger specific agent actions. With such flexible structure, agents can switch strategies during the simulation, or execute multiple strategies at different points in time.

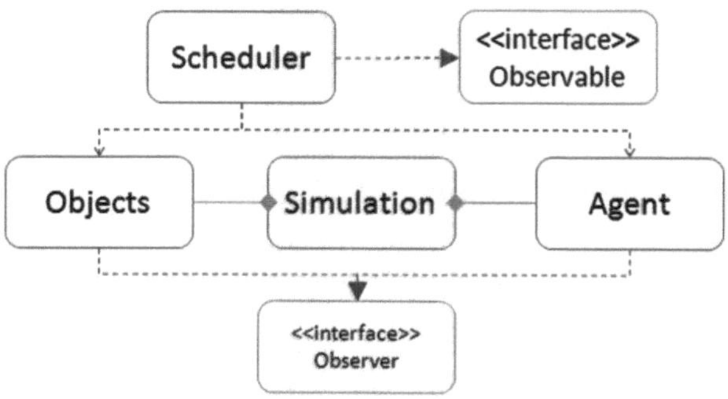

Fig. 2 Simplified UML class diagram of an observer design pattern

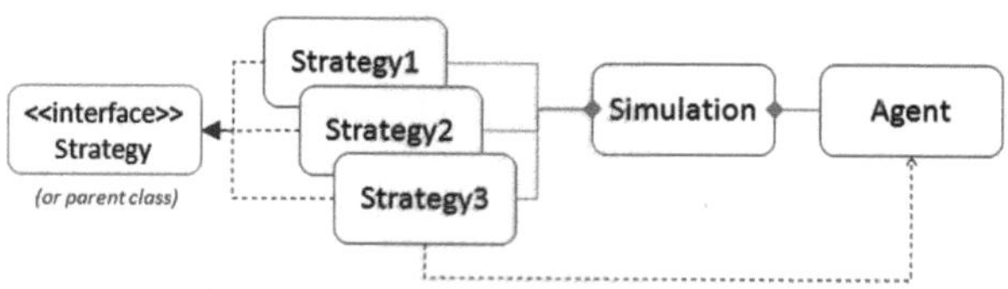

Fig. 3 Simplified UML of a strategy design pattern

Other Design Patterns Among other design patterns useful in ABM, the singleton pattern can ensure an unique instance of specific classes, such as the environment class or the scheduler class. The factory pattern helps to cope with heterogeneous model-specific objects such as transactions (risky or risk-free), etc. Finally, the adapter pattern supports the development of the most complicated ABM structures.

3.2.3 Usage of Realistic Data Encapsulation

Every object does not need an access to every simulation parameter. Interestingly, encapsulation in ABM brings the opportunity to mimic real world structures of private and public information. Let us suppose a model with a complex bank agent class, which burden has been delegated to "bank divisions" classes. One could implement a realistic data encapsulation: only the accounting division could access to the bank books, rendering the model dynamics and information management even more realistic and understandable without increasing its complexity.

3.2.4 Agent Lifecycle Management

The need of multiple "runs" in a simulation should be implemented through OO construction and destruction of objects. Keeping the same environment and agents objects, reinitialized at each run, is unnecessarily complex and can generate code or memory errors. For each run, the non-technical objects should be re-created, fully benefiting from the constructor feature.

4 Concrete Application of the Methodology

In this section, we will present increasingly advanced implementations of ABM from an abstract perspective, then applied to a concrete case: the Black Rhino (BR) model (Georg 2013), from the assumptions to the final implementation structure.

4.1 The Black Rhino Assumptions and Model

The BR model is an agent-based financial model designed to study the contagion of systemic risk among connected banks. Discussing the theoretical scope of the model is beyond the topic of this paper. Here below, we list its hypothesis:

- Banks receive a random amount of deposit at each period.
- Banks decide on the unconstrained optimal investment volume on the basis of a self risk assessment. There are no learning dynamics.

- After settlements of interests, loans and deposits cash flows, banks are able to invest excess in liquidity on the interbank market, then in new investments if their optimal volume is not reached, then in central bank loans.
- Banks enduring a default of cash will borrow on the interbank market, then from the central bank. If the shortage persists, they will sell assets. Ultimately, they bankrupt. A bank under bankruptcy is removed from a simulation run.
- Investments can individually randomly default at a random time.
- Random exogenous shocks can decrease a bank equity or liquidity.
- Interest rates are paid or received on investments, interbank loans, central bank loans, and deposits.
- Banks are connected in a network, and can only make interbank trades with neighbors. The network topology is constant during the simulation.
- Simulation parameters can vary during a run, at pre-determined times.

The goal of the model will be the analysis of banks states after a defined among of timesteps, for specific network topologies, initial setups and exogenous shocks.

Translating the hypothesis in a model is straightforward: we identify banks as the agents given their decision making regarding the investment volume, their interconnectedness, and the defined set of information and actions at their disposal. The system will evolve at each timestep, on a yearly frequency if the model is calibrated on yearly data.

Given that banks must settle interests, loans and deposits cash flows before interacting, their sequence of actions must be split in phases. Therefore, banks will have only one strategy which phases will be:

- Pre-interbank processing: Settlement of interests, loans and deposits—Computation of the optimal investment volume—Definition of liquidity needs
- Interbank processing: Obtaining or providing liquidity to neighbors
- Post-interbank processing: If default in liquidity, obtaining cash from central bank—If default in liquidity, obtaining cash from assets sales—If default in liquidity, settle accounts and remove bank from the simulation.

4.2 The Original Implementation

The rationale of the initial implementation (Fig. 4) is as follows: the simulation "Black Rhino" creates an environment which sets up the banks, the network of banks, the parameters list and the state, an object that always contains the up-to-date parameters values. The simulation also creates a runner and the measurements, the former executing the requested amount of simulation and the latter retrieving results and redirecting outputs in an external file. Finally, the updater is created by the runner. It orders each bank to perform its routines at each timestep, and eventually applies exogenous shocks to the system.

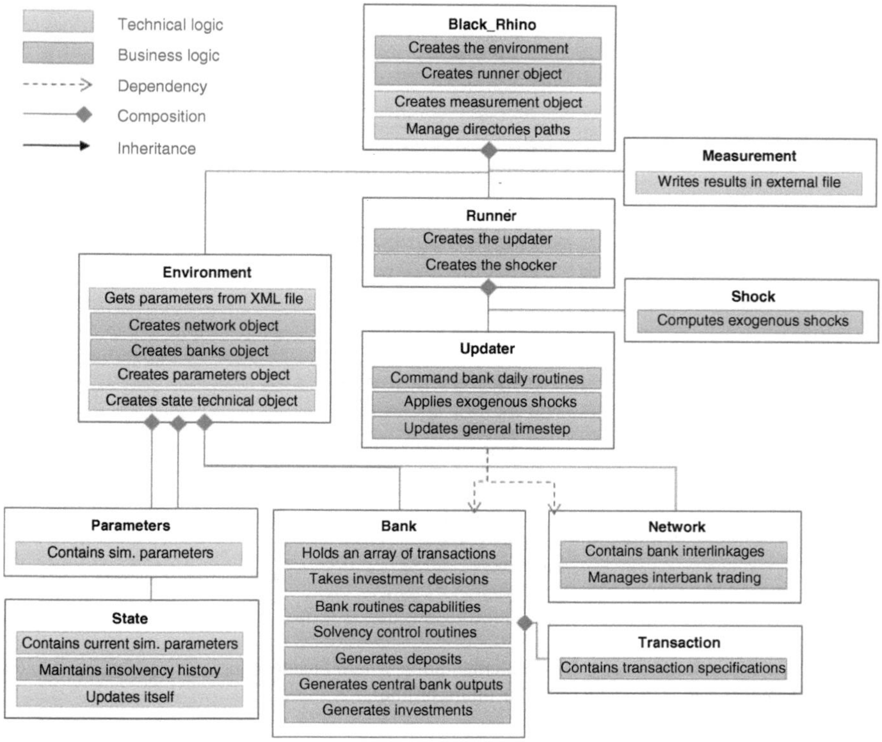

Fig. 4 The simplified UML class diagram of the original Black Rhino model

4.3 A First Step: Separating the Business Logic

Having the technical objects separated from the business ones is a trivial yet important step forward. The separation requires agents to be completely dependent of a world class (Fig. 5), sometimes called environment or economy or market, which stands as a frontier between the business and the technical logic. Its main purpose is the creation of the agents and the activation of their internal routines. The implementation also automatically provides the automated management of agents' lifecycle: they are constructed and destroyed upon construction and destruction of the world.

The Case of Black Rhino We applied this recommendation to the Black Rhino model and obtained a first structure, illustrated in Fig. 5. Here, when the simulation instantiates an environment, the subsequent instantiations are exclusively business logic-centric. Input and output dynamics and the initialization of the parameters have a dedicated technical area. Also, the network becomes the bank container, instead of being an auxiliary informative structure.

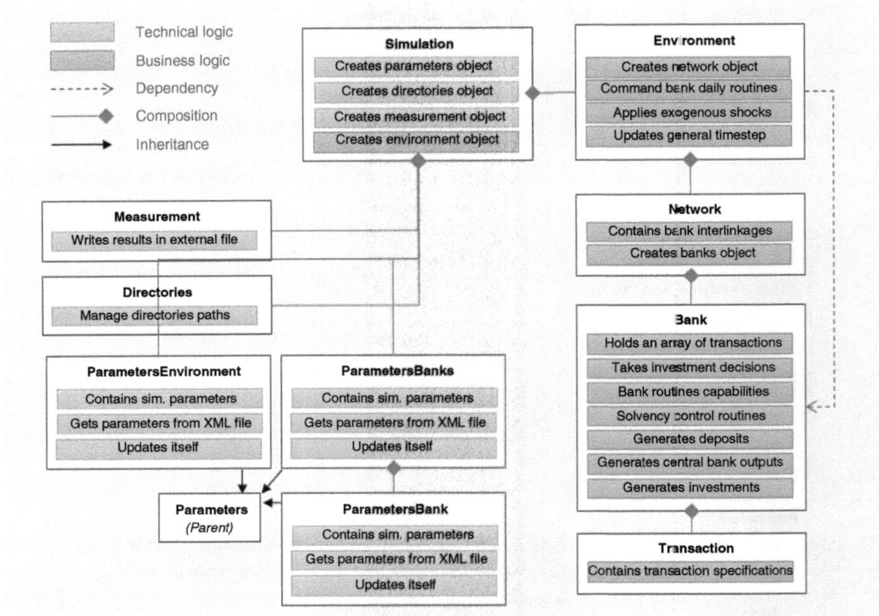

Fig. 5 UML Class diagram of a BR implementation where business logic (in *green*) is isolated from the technical logic (in *orange*) (Color figure online)

Such implementation presents many advantages:

- Robustness: updating the model is easier and safer
- Simplicity: researchers are able to focus on the business logic, being in this case an independent part of the model
- Clarity: technical objects are persistent, while business object are refreshed at each simulation run. Since this structure does separate both kinds, an elegant way to refresh the business part is the re-instantiate the objects, hence using the construction/destruction feature of the object-oriented paradigm.

Despite structural improvements, several shortcomings entice us to pursue the planning phase:

- The implementation does not allows agents to use multiple or varying decision strategies.
- The agents are executing routines that are, in the real world, more likely to be operated by external entities (e.g. a central bank).
- The model dynamics are nearly exclusively found in the agents routines, which hampers the understandability.

4.4 The Second Step: An Advanced ABM Framework

Implementing the strategy pattern in case of heterogeneous agents behaviors or learning capabilities is a must. We extract the routines of decision making and learning of the agent inside separate classes. It clarifies the content of the business classes and enables some adaptive capabilities, making it easier to modify the behavior of agents.

We also integrate the observer pattern which unloads the world from some dynamics, relegating it to the sole role of creating and destroying the agents and the adjuvants.

Finally, we extract some business logic into the new objects, with a business meaning but no decision making. This is a part of our modeling efforts to comply with the structure of the real world. Indeed: special treatments provided to agents such as checks for bankruptcy or food production can be handled by some regulator or food production firm classes, and interact with agents in the same world. The idea is to encapsulate those dynamics handled initially by the world class into specific classes that make sense in comparison with the real world (Fig. 6).

This next step brings even more benefits:

- Robustness: we enabled modularity in agents' strategies and learning and we can now easily add, remove or modify the interacting objects such as regulators, etc.
- Simplicity: closely complying with the ABM classical framework, the implementation can now be handled by practitioners with ease. We also facilitated data collection within the newly created interacting objects (e.g. total households' deposits during the last timestep).

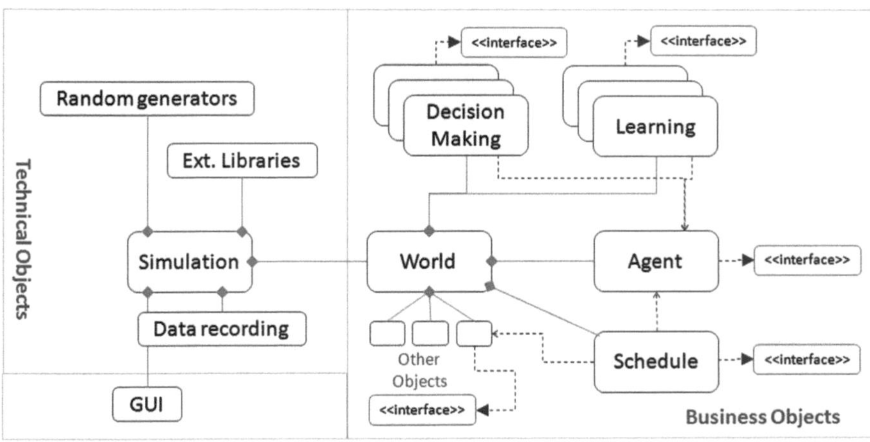

Fig. 6 UML Class diagram of an ABM implementation with extended evolution capabilities, including common design patterns

- Clarity: we improved the clarity without increasing the complexity. The business objects structures and capabilities now match closely their real world counterparts.

The Case of Black Rhino First, we extract the timestep management from the environment and encapsulate it in a "Scheduler"; a very common name for the observer pattern in ABM simulation (Luke et al. 2003). Then, we extract the three phases presented in the model specification section from the bank to a strategy entity called "BankStrategy" to obtain the strategy pattern. Additional BankStrategies can be created and assigned to banks initially or under specific conditions. Since a scheduler, a bank and a BankStrategy are respectively supposed to "notify", "update" and "process", we append the subsequent interfaces to enrich the implementation.

Finally, the last shortcomings are addressed by creating entities interacting with agents while not being agents themselves since they do not take any decisions. Banks were provided capabilities of a central bank, of households, of firms and of a legislator. Encapsulating those dynamics in related entities provide the next structure (Fig. 7).

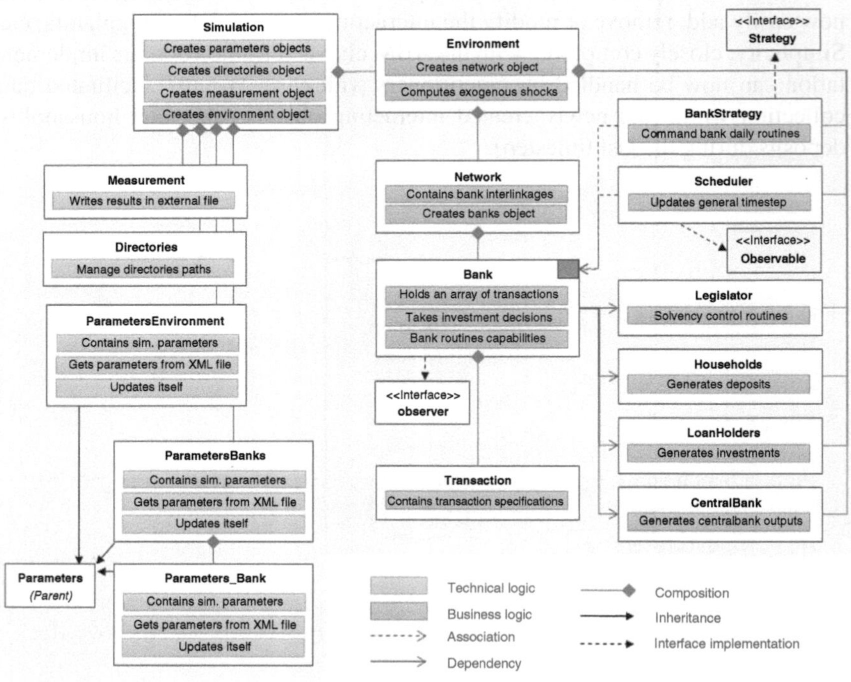

Fig. 7 UML Class diagram of a BR implementation closely related to the ABM framework

Within the scope of this case study, halting the development of the structure at this point would be reasonable, since it can be generalized to the vast majority of economics and finance-related ABMs. The following sections explore additional refinements only fitting more specific cases.

5 Case-Specific Discussions

5.1 Implementing Capabilities

In specific modeling cases, agents and some other simulated objects could have common capabilities. For example, banks are able to hold accounting book objects, which is also the case of the central bank. In such case, the bookkeeping system can be separated as a module, inherited by any capable object of the simulation, further increasing the clarity and the realism of the model (Fig. 8).

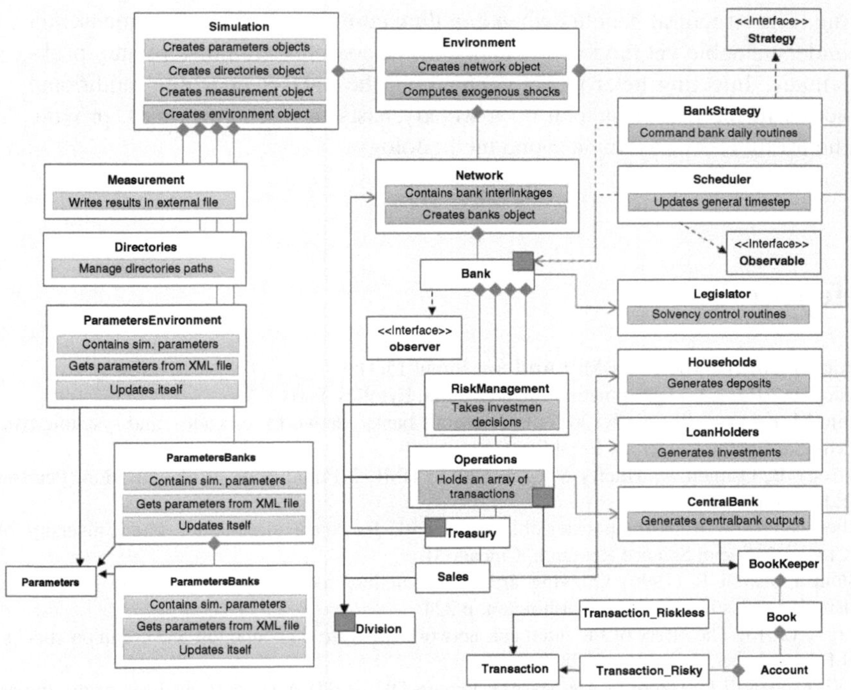

Fig. 8 UML Class diagram of a BR implementation where Bank is split into comprehensive units, sharing capabilities with other objects

5.2 *Object-Oriented Trade-Off*

While implementing ABMs according to the OO paradigm, one should keep the classes structure bound to the real world counterpart. If the agent is a consumer, its decision-making and other routines should be included in a consumer class. However, when the models exigencies are complex, it can be wise to spread the agent routines in internal classes wisely. In the case of Black Rhino, we created bank divisions: risk management, operations, sales and treasury, each one having its specific behavior and access to other objects. As seen in Fig. 8, the operations division, for example, has a unique access to the central bank.

This subsection name suggests a trade-off, since it clarifies the main class and improves realism, at the cost of a complex and rigid structure.

Conclusion
Agent-Based models are a very fertile soil for the establishment of structured implementations. Given the synergy between OO, UML and ABM paradigms, researchers can grasp the importance of carefully planning before implementing. The structural benefits when handling complex models are immense: an understandable yet flexible model tacking a complex multidisciplinary problematic. Injecting heterogeneity, changing the dataset or adding additional actors in the model happen on a weekly basis in research groups, proving the usefulness of having a sound methodology.

References

Bersini H (2012) UML for ABM. J Artif Soc Simul 15(1):9

Bersini H (2013) La programmation orientee objet. Eyrolles, Paris

Bluhm M, Krahnen JP, Faia E (2012) Endogenous banks' networks, cascades, and systemic risk. Mimeo, Goethe University Frankfurt

Charroux B, Osmani A, Thierry-Mieg Y (2013) UML 2 Pratique de la modélisation. Pearson Education, Paris

Collier N (2003) Repast: an extensible framework for agent simulation, The University of Chicago's Social Science Research, Chicago IL

Epstein J, Axtell R (1996) Growing artificial societies: social science from the bottom up. Brookings Institution Press, Washington, p 224

Georg C (2013) The effect of the interbank network structure on contagion and common shocks. J Bank Financ 37(7):2216–2228

Iori G, Caldarelli G, Gabbi G, De Masi G, Precup OV (2008) A network analysis of the Italian overnight money market. J Econ Dyn Control 32:259–278

Luke S, Catali-Balan G, Liviu P, Cioffi-Revilla C, Paus S (2003) MASON: a java multi-agent simulation library. In: Proceedings of the agent 2003 conference

Building Artificial Economies: From Aggregate Data to Experimental Microstructure. A Methodological Survey

Gianfranco Giulioni, Paola D'Orazio, Edgardo Bucciarelli, and Marcello Silvestri

1 Introduction

Agent-Based Modeling has been contributing to the renewal of economic methodology by relying on the algorithmic approach in order to model economic phenomena (Velupillai 2011). This approach allows to focus on the dynamic nature of the economy and for explicitly introducing *heterogeneity* and *interaction* into the models (Kirman 1992).

From the mid-1990s, researchers started studying results collected from human subjects' experiments by using agent-based models (ABMs). Brian Arthur (1991) was among the firsts in exploring the idea of calibrating an algorithm to reproduce human behavior. He called the attention on the need to go beyond the assumption of *rationality* by suggesting some ways to model economic choices by means of an algorithm that was *"tuned to choose actions in an iterated choice situation the way humans would* " (Arthur 1991, p. 354). To calibrate the algorithm in a way that could be defined as a "good indication" of human behavior, he used the results of an experiment performed in 1952–1953 by Robillard at Harvard University. Simulations results, and related tests of fitness, were very striking in that they showed that Arthur's automaton was able to replicate behaviors observed in the experimental laboratory in different choice problems than those for which it was calibrated. As reported and discussed in Dawid (1996), Arifovic (2000) and

G. Giulioni (✉) • P. D'Orazio • E. Bucciarelli • M. Silvestri
Department of Philosophical, Pedagogical and Economic-Quantitative Sciences, "G. D'Annunzio" University, Viale Pindaro 42, 65127 Pescara, Italy

Research Group for Experimental Microfoundations of Macroeconomics (GEMM), Pescara, Italy
e-mail: g.giulioni@unich.it; paola.dorazio@unich.it; e.bucciarelli@unich.it; marcello.silvestri@yahoo.it

F. Amblard et al. (eds.), *Advances in Artificial Economics*, Lecture Notes in Economics and Mathematical Systems 676, DOI 10.1007/978-3-319-09578-3_6

Duffy (2006), starting from Arthur's contribution, *Evolutionary Algorithms* (EA) (in particular the *Genetic Algorithm*, GA) have been extensively used in Economics.

A closer look to the works published in the 1990s and early 2000s suggests that to calibrate artificial agents and artificial economies, researchers rely mainly on *aggregate data*. This implies that ABMs have been focusing more on the *social* rather than on the *individual* level.

This paper surveys the recent literature (2005–2013) highlighting two ongoing tendencies:

- Researchers have been focusing more on the *micro empirical structure* than on the aggregate framework. The main advantage of this change lies in the possibility of accounting for a "complete" *heterogeneity* in modeling agents' behavior. The type of interaction between Experimental Economics (EE) and Agent-based computational economics (ACE) implied by this tendency is not new and, in this sense, this paper can be viewed as an update of Duffy (2006). We will refer to this type of interaction as the EE/ACE relationship.
- Experimental economists are gradually acknowledging the potential benefits that experiments run with human subjects can gain from simulations in an artificial environment. This in turn implies a kind of "reverse" relationship, i.e., ACE/EE.

The present paper is an extension of D'Orazio and Silvestri (2014) and aims at building an up-to-date overview of the methodological improvements and developments in the EE/ACE and ACE/EE interactions witnessed in the last years. The remainder of the paper is organized as follows. After briefly recalling the early steps made in the methodological interaction EE/ACE, Sect. 2 discusses the focus shift from *single-population* towards *multi-population* algorithm. We will highlight how this process emphasizes the problems related to data availability and the calibration of artificial agents. Section 3 discusses how Experimental Economics gained importance in shedding light both on the micro and the macro economic theory. Furthermore, we try to emphasize the methodological improvements and the theoretical results of a set of papers which rely on the interaction between the experimental method and evolutionary techniques. Section 4 discusses the way in which human subjects' experiments can benefit from the insights of ACE simulations (ACE/EE relationship). Section "Conclusions" closes the paper.

2 Multi-Population Versus Single-Population Algorithms

Most works in Computational Economics have generally made use of *single population algorithms*.[1] Recently, ABM researchers are focusing more on the *individual-based approach* trying to replicate agents' behavior at the individual level, thus accounting for a more *complete* heterogeneity among agents. In this

[1]It is worth recalling the counterintuitive nature of these two definitions. *Multi-population* algorithm refers to those EA in which researchers calibrate individual agents. *Single-population* algorithms refers to the social level. See, among others, Holland and Miller (1991).

way, computational works have both a micro and a macro "flavour". In particular, according to methodological applications surveyed in this paper, we argue that by adopting the *individual-based* perspective, the *self-referential* nature of macroeconomic models can be improved (Arifovic 2000).

2.1 Artificial Agents Behavioral Rules: From Aggregate Data to an Empirical Microstructure

Duffy (2006) explores three main ways in which the combination of the agent-based approach and human subjects experiments has been used. He stresses that this combination allows for a useful comparison between data collected from human experiments with those resulting from computer simulations. According to his review, the first approach used in the economic community is the *"zero-intelligent" agent* approach; it is based on the Gode and Sunder seminal work and consists of an agent-based model with very low rationality (Gode and Sunder 1991). A second approach employs more sophisticated individual behaviors, ranging from simple stimulus-response learning to more complicated belief-based learning approaches. A third, more recent approach, considers a more complicated individual learning process by means of classifier systems or genetic algorithms (GA).

A closer look at the literature on the EE/ACE interplay shows that most works—both in Economics and Finance—make use of *aggregate* data in order to calibrate agents. As shown in the Arifovic (1996) seminal paper, Agent-based models which compare human and artificial agents behaviors have indeed been used to *mirror* human subjects macroeconomic experiments results. The *aggregate* variables (prices, exchange rates, etc.) are firstly generated in the experimental laboratory by observing human subjects' behavior. Then, artificial agents are calibrated in order to replicate experimental results (see Dawid 1996; Arifovic 2000, for a comprehensive survey). At the same time, *heterogeneity* has been considered a key feature in that this enables the exploration of the possible emergent complexity at the micro level (Kirman 2006) both in Agent-based computational models and in Agent-based analytically tractable models (Hommes 2006). Indeed, the ABM literature has always acknowledged the importance of *heterogeneous* agents in that this enables the exploration of the possible *emergent complexity* at the micro level. Nevertheless, artificial agents are usually considered as "equally smart" (Chen and Wang 2011) while the building of behaviorally richer models seems to have drawn researchers' attention only recently.

The building of the empirical microstructure in an ABM suffers however from the paucity of *individual data* compared to aggregate data; the existing empirical datasets are not detailed enough to allow for an accurate identification of agents' behaviors. Economists are thus tangled up with many "degree of freedom" in the choice of the microeconomic model.

By observing human subjects behaviors in the experimental laboratory, Experimental Economics could help in bridging this gap. Recent works show that the use of *experimental micro data* to calibrate artificial agents is among the most important innovations in that, *ceteris paribus* they improve the *self-referential* nature of macroeconomic models.

3 Exploring Economic Agents: Experimental Insights in ABM

3.1 Micro and Macro Experiments: Do We Need to Set Them Apart?

According to Smith (1982) the experimental literature can be divided between *methodological* and *functional* experiments.

Methodological experiments shed light on the functioning of microeconomic systems and they have been performed to address basic and evident objections to the standard *microfoundations* of macroeconomic models, i.e., to examine market predictions, individuals rationality and agents preferences. Experimental results stemming from this vast body of existing literature provide evidence of a mismatch between the predictions of conventional theory (e.g., the standard optimization assumption) and the actual behavior of individuals (Tversky and Kahneman 1974).

Functional experiments aim at establishing laws of behavior on both the theoretical and empirical side. They are related to the institutional side of an economic model, i.e., trading rules, auction designs and matching mechanisms.

The description sketched above suggests that experiments in Economics have been conducted more on the "*micro-micro*" side rather than on the "*micro-macro*" side. According to Sargent (2000) this tendency may be due to the fact that "the choices confronting artificial agents within even one of the simpler recursive competitive equilibria used in macroeconomics are very complicated relative to the settings with which experimentalists usually confront subjects" (Sargent 2000, footnote 45, page 27) However, as extensively surveyed by Duffy (2008), researchers are increasingly considering macroeconomic models and test them in the lab: "[...] The main insights from macroeconomic experiments include: 1) an assessment of the micro-assumptions underlying macroeconomic models, 2) a better understanding of the dynamics of forward-looking expectations, 3) a mean of resolving equilibrium selection (*coordination*) problems in environments with multiple equilibria, 4) validation of macroeconomic model predictions for which the relevant field data are not available and 5) the impact of various macroeconomic institutions and policy interventions on individual behavior" (Duffy 2008, p. 3).

In our view, this debate suggests that a sharp distinction between micro and macro experiments cannot be drawn because of the distinctive features of modern macroeconomic models, i.e. their explicit microfoundations and the related representative agent assumption.

3.2 *Fitting Evolutionary Models: The Experimental Micro Level*

In the last 20 years, theoretical models of *heterogenous bounded rational agents* (Hommes 2006) and agent-based financial models have literally blossomed for they provide a more complete, flexible and consistent representation of the actual market mechanisms (see LeBaron 2000; Lebaron 2001, for comprehensive surveys).

The main concerns of agent-based financial markets modelers have been (a) a reasonable representation of a financial trading situation in the model and (b) the quantitative replication of the most important features of real markets via a reasonable calibration. By considering markets as large aggregations of agents with heterogenous beliefs, agent-based markets allow for the understanding of some important empirical puzzles found in standard representative agent financial models, such as the equity premium puzzle, and to study important issues such as the volatility persistence and the use and performance of different trading rules.

In these models *agents* have generally been considered as simple bounded rational actors[2] and have been *estimated* with empirical or survey data. However, it is important to recall that *economic data* are available more at the aggregate/macroeconomic level, while *financial data* are easily available at the disaggregate level, from annual to minute by minute. Indeed, the use of heterogeneous agents is not new in finance studies as confirmed by the wide literature on heterogeneous agents rational expectations models. What is new in computational frameworks surveyed in this paper is that they are trying to cope with the problem of *complex heterogeneity* which makes models analytically intractable. Furthermore, our survey suggests that if researchers are willing to build more accurate models of the interaction of heterogenous agents, they need different tools (see Chen 2012, for a comprehensive survey on concepts and design of agents in ABM).

In order to cope with this new need, the economics community has indeed engaged in a new research path, starting to use experiments to collect data to calibrate the micro level (Giulioni et al. 2014) and to study agents behaviors in a "constrained" and controlled environment. Moreover, considering that the *learnability* of rational expectation equilibria depends on the structure of the feedback system between individual expectations and the economic environment, experiments are of great help in this research strand in that they allow to go deeper in the *feedback mechanism*. By means of the experimental method is thus possible to control for confounding factors and observe specific issues in order to study adaptive learning and its stability.

[2] As pointed out by LeBaron (2000), any modeler should take into account some basic issues about agents' design in that results are inevitably influenced by the learning methods agents have been endowed with. Moreover, he argued that the problem of the bounded memory perspective on past information (i.e., the time horizon) and that the quantity of data, i.e., information, agents should use to take their decisions are crucial in financial models.

By adopting the "*macro* experimental perspective", several experiments have been performed on agents' *learning* process in complex macroeconomic systems. Cars Hommes was among the firsts in exploring the behavioral space, namely agents' expectations heterogeneity, at the experimental level. His research has been motivated by the joint search for a *large* computational heterogeneous agent-based model able to capture the stylized facts as closely as possible and *the simplest* behavioral heterogeneous agents model with a plausible behavioral story at the micro level.

The first results of lab experiments with human subjects that find support for *heterogeneity* among expectations are reported in Hommes (2007). These experiments are labeled as *learning-to-forecast* experiments (LtFEs) because subjects are asked to be *only* forecasters (not producers, nor traders) of the price of some commodity knowing that their earnings are inversely proportional to their forecast error. They have *qualitative* information about the market: they *do not know* other participants' forecasts, the exact market equilibrium equation, the exact demand and supply schedules and the exact number of other demanders and/or suppliers in the market but they *know* past prices and their own past forecasts and earnings. In particular, they know that the *price* is an aggregation of individual forecasts, derived from *equilibrium* and are able to infer the type of expectations feedback: positive or negative. *Positive* (*negative*) feedback means that an increase of (average) individual forecasts leads to a *higher* (*lower*) market equilibrium price. *Positive feedback* mechanisms play a key role in speculative asset markets: higher market expectations lead to an increase of speculative demand and thus to an increase of the realized price of that asset. *Negative feedback* may be dominant in supply driven commodity markets, where an increase in expected prices leads to higher production and thus to a lower realized market price.

After experiments are run with different market settings (stable, unstable, strongly unstable), experimental results are used to validate expectation hypotheses and learning models. By combining the experimental method and evolutionary techniques, Hommes provided evidence for the importance of heterogeneity in a theory of expectations: by means of a simple *heuristics switching model* it is possible to fit different behaviors collected from LtFEs. These results are indeed crucial for economic theory in that they clearly demonstrate that the *rational expectation hypothesis* occurs only in stable markets (Hommes 2011).[3]

By adopting the methodology discussed above, Assenza et al. (2013) presents (early) results and monetary policy implications of a *heterogenous expectations switching model* where the aggregate variables depend on individual forecasts of two variables, namely, the *output gap* and *inflation*. The *learning-to-forecast* process is observed in the lab and results collected are used to fit a simple evolutionary New Keynesian macroeconomic model. The authors argue that the model helps in drawing some important policy considerations: compared to the standard rational

[3]More details on those results and the *learning to forecast* experimental literature can be found also in Bao et al. (2013).

expectation model, a simple model based on evolutionary switching which is able to account for heterogenous expectations can generate persistent deviations from the steady state. In other words, the convergence to the desired monetary policy target can be slow if we reject the homogenous adaptive learning rule.

In a more recent work Hommes and Lux (2013) retake Hommes (2007) experimental results and, in an attempt to go beyond the simple heuristics switching model, propose a GA as a way to model *individual expectations* and explain aggregate market phenomena. They started by reproducing as close as possible the design of experiments in the GA. In particular, they initialized it with the same number of agents, the same parameters of demand and supply functions as in the experiment and choose the fitness function identical to the payoff function used in the lab. They perform simulations with individual forecasting rules in different treatments (environments) and increasing the number of interacting agents (from 6 to 30). Simulations results show that artificial agents' interaction is able to explain all stylized facts observed in the aggregate price *simultaneously* and across treatments, in a way that is consistent with theoretical predictions of heterogeneous expectations endogenous selection. Furthermore, they stress the relevance of running GA-simulations with experimental data, highlight the importance of the cross-fertilization between experimental economics and evolutionary techniques. This point is discussed in details in Sect. 4.

Anufriev et al. (2013) engage in the search of better microfoundations for their model by using GA-based individual learning. Compared to other similar works considered so far, they use many different experimental settings; this makes the model useful to investigate situations in which heterogeneous price expectations have important consequences for market efficiency and dynamics. Furthermore, by means of Auxiliary Particle Filter the authors claim that the model explains the *individual*, not just the aggregate results of the LtF experiments, thus going beyond the standard "aggregate" focus of agent-based models.

4 Matching Pairs: ACE Insights into Human Subjects Experiments

The rapidly expanding research strand based on the interaction EE/ACE suggests that they are "natural allies" in that they help each other in coping with their *external validity* shortcomings (Duffy 2006). They indeed complement each other: EE helps ACE in dealing with its "degree of freedom" problem and ACE helps EE in controlling and providing benchmarks for experimental subjects' behavior. As discussed in the introduction, GAs have been acknowledged as realistic models of human cognition because they provide robust search in complex spaces.

A closer look at the literature shows that several studies have also been conducted in the reverse direction, i.e., human subject experiments conducted in light of ACE results. Casari's (2004) article explores this issue by showing that simulations with GA allow to: (a) make comparisons with experimental data and (b) make predictions about the effects of different experimental designs. The novelty of Casari's results lies mainly in the latter. He found that agents with identical goals and identical, although limited, levels of rationality behave in *different* manners, i.e., GA generates individual different patterns. Changes in the experimental design, as e.g. the restriction of the agents' strategy space, are then explored; the resulting predictions are also supported by experimental results. This particular use of GA simulations has been extended by Casari himself (Casari 2008) and further studied by Chen et al. (2008) and Chen and Yu (2011).

Hommes and Lux (2013) call the attention on the methodological advantages of performing GA-simulations calibrated with experimental data. After their simulations, they are able to draw some important insights on the collected results. First of all they claim for the need of additional laboratory experiments in order to get more information about the number of strategies used by subjects: more experiments will help in understanding which forces dominate in different settings. Furthermore, using a GA to expand the number of interacting agents make them inferring that a number greater than 30 subjects has relatively little impact on aggregate price behaviors. This is a potential important result for the experimental method, in particular for the debate on the external validity of experiments run with a small (say, less than 30 subjects) subject pool.

Conclusions

The goal of this survey is twofold. First, it highlights the main features which characterized the methodology based on the interplay between EE and ACE in the last 5–6 years. Second, it provides some points for further discussion on methodological improvements.

We report on the methodological debate about the *trade-off* between small (in terms of number of parameters and variables) and "large" behavioral ABMs and present and discuss (a) a recent strand of the literature which is working for building richer behavioral models and (b) some "alternative" uses of artificial simulations aimed at improving economic experiments.

The literature surveyed in this paper calls into question further improvements in the application of GA in Economics and Finance. Dawid and Dermietzel (2006) and Waltman et al. (2011) highlighted this need by claiming that all elements of an EA should have *meaningful economic interpretation*. Besides the concern about the soundness of *population size* in economic environments (discussed in Sect. 2), the main argument of the paper is that, as the method evolves and the literature grows, it is important to shed light on some crucial methodological issues. Some possible improvements concern the use of more recent computational techniques developed in the fields of artificial intelligence, heuristic optimization and so on. Additionally, although the debate on these issues in the ACE community is not new, a

(continued)

thorough reconsideration of the convenience of the *binary encoding* versus *real-coded* chromosomes (Herrera et al. 1998) of agents' strategies is needed, especially in the light of recent methodological developments highlighted in our paper. Furthermore, this review calls the attention on the use of the *selection* and *crossover* operators in GA.

Another relevant issue on which the paper calls attention concerns the policy analyses—as well as the policy recommendations—that economists and policy makers can draw from an ABM built on the interaction between EE and ACE. Farmer and Foley (2009) addressed this issue by claiming that by using ABMs "*policy makers can thus simulate an artificial economy under different policy scenarios and quantitatively explore their consequences*". We maintain that the EE/ACE and ACE/EE relationships can provide sounder microfoundations to macroeconomic ABM that will in turn further improve the reliability of policy recommendations.

References

Anufriev M, Hommes C, Makarewicz T (2013) Learning to forecast with genetic algorithm. Tech. rep.

Arifovic J (1996) The behavior of the exchange rate in the genetic algorithm and experimental economies. J Polit Econ 104(3):510–541

Arifovic J (2000) Evolutionary algorithms in macroeconomic models. Macroecon Dyn 4:373–414

Arthur WB (1991) Designing economic agents that act like human agents: a behavioral approach to bounded rationality. Am Econ Rev Pap Proc 81(2):353–359

Assenza T, Heemeijer P, Hommes C, Massaro D (2013) Individual Expectations and Aggregate Macro Behavior. Tinbergen Institute Discussion Papers 13-016/II, Tinbergen Institute

Bao T, Duffy J, Hommes C (2013) Learning, forecasting and optimizing: an experimental study. Eur Econ Rev 61(C):186–204

Casari M (2004) Can genetic algorithms explain experimental anomalies? Comput Econ 24(3):257–275

Casari M (2008) Markets in equilibrium with firms out of equilibrium: a simulation study. J Econ Behav Organ 65(2):261–276

Chen SH (2012) Varieties of agents in agent-based computational economics: a historical and an interdisciplinary perspective. J Econ Dyn Control 36(1):1–25

Chen SH, Wang SG (2011) Emergent complexity in agent-based computational economics. J Econ Surv 25(3):527–546

Chen SH, Yu T (2011) Agents learned, but do we? Knowledge discovery using the agent-based double auction markets. Front Electric Electron Eng China 6:159–170

Chen SH, Zeng RJ, Yu T (2008) Co-evolving trading strategies to analyze bounded rationality in double auction markets. In: Genetic programming, theory and practice VI. Springer, Heidelberg, pp 195–213

Dawid H (1996) Adaptive learning by genetic algorithms: analytical results and applications to economic models. Springer, New York

Dawid H, Dermietzel J (2006) How robust is the equal split norm? Responsive strategies, selection mechanisms and the need for economic interpretation of simulation parameters. Comput. Econ. 28(4):371–397

D'Orazio P, Silvestri M (2014) The empirical microstructure of agent-based models: recent trends in the interplay between ACE and Experimental Economics. In: Proc. 11th Int. Symp. Distrib. Comput. Artif. Intell., pp 1–6

Duffy J (2006) Agent-based models and human subject experiments. In: Tesfatsion L, Judd KL (eds) Handbook of computational economics, chap 19, vol 2. Elsevier, Amsterdam, pp 949–1011

Duffy J (2008) Macroeconomics: a survey of laboratory research. Working Papers 334, University of Pittsburgh, Department of Economics

Farmer JD, Foley D (2009) The economy needs agent-based modeling. Nature 460:685–686

Giulioni G, Bucciarelli E, Silvestri M, D'Orazio P (2014) Avatar-based macroeconomics - experimental insights into artificial agents behavior. In: Proc. 6th Int. Conf. Agents Artif. Intell. SCITEPRESS - Science and Technology Publications, pp 272–277. doi:10.5220/0004917902720277. http://www.scitepress.org/DigitalLibrary/Link.aspx?doi=10.5220/0004917902720277

Gode D, Sunder S (1991) Allocative efficiency of markets with zero intelligence (z1) traders: market as a partial substitute for individual rationality. Gsia working papers, Carnegie Mellon University, Tepper School of Business

Herrera F, Lozano M, Verdegay JL (1998) Tackling real-coded genetic algorithms: operators and tools for behavioural analysis. Artif Intell Rev 12(4):265–319. doi:10.1023/A:1006504901164

Holland JH, Miller JH (1991) Artificial adaptive agents in economic theory. Am Econ Rev 81:365–370

Hommes C (2006) Heterogeneous agent models in economics and finance. In: Tesfatsion L, Judd KL (eds) Handbook of computational economics, chap 23, vol 2. Elsevier, Amsterdam, pp 1109–1186

Hommes C (2007) Bounded rationality and learning in complex markets. CeNDEF Working Papers 07-01, Universiteit van Amsterdam, Center for Nonlinear Dynamics in Economics and Finance

Hommes C (2011) The heterogeneous expectations hypothesis: some evidence from the lab. J Econ Dyn Control 35(1):1–24

Hommes C, Lux T (2013) Individual expectations and aggregate behavior in learning-to-forecast experiments. Macroecon Dyn 17:373–401

Kirman AP (1992) Whom or what does the representative individual represent. J Econ Perspect 6:117–136

Kirman A (2006) Heterogeneity in economics. J Econ Interact Coord 1(1):89–117. doi:10.1007/s11403-006-0005-8

LeBaron B (2000) Agent-based computational finance: suggested readings and early research. J Econ Dyn Control 24(5–7):679–702

Lebaron B (2001) A builder's guide to agent based financial markets. Quant Financ 1:1–2

Sargent TJ (2000) Evolution and intelligent design. Am Econ Rev 98(1):5–37

Smith VL (1982) Microeconomic systems as an experimental science. Am Econ Rev 72(5):923–55

Tversky A, Kahneman D (1974) Judgment under uncertainty: heuristics and biases. Science 185:1124–1131

Velupillai KV (2011) Towards an algorithmic revolution in economic theory. J Econ Surv 25(3):401–430

Waltman L, Eck N, Dekker R, Kaymak U (2011) Economic modeling using evolutionary algorithms: the effect of a binary encoding of strategies. J Evol Econ 21(5):737–756

Spontaneous Segregation of Agents Across Double Auction Markets

Aleksandra Alorić, Peter Sollich, and Peter McBurney

1 Introduction

Adam Smith, in his *The Wealth of Nations* said that the concept of economic growth is deeply rooted in the division of labour. This primarily relates to the specialization of the labour force, where narrowing expertise allows better exploitation. Contemporary examples of such specialization include, e.g., airline companies: some specialize in first class and business flights, while others provide mainly low cost flights. The paper (Nagarajan et al. 1995) reports segmentation phenomena in the informal credit market in the Philippines, where lenders who specialize in trading make loans mainly to large and asset-rich farmers, while others lend more to small farmers and landless labourers.

It can be argued that the space of customers is already segmented, and that the role of an efficient merchant is to find and adapt to niches in this customer space (see for example Robinson et al. 2012). However, here we want to explore the possibility of spontaneous segregation of initially *homogeneous* traders. This work was motivated by observations from *the CAT Market Design Tournament* (Cai et al. 2009) where competitors were invited to submit market mechanisms for a population of traders provided by the tournament organizers. It was observed that by co-adaptation of markets and traders the system evolved to a segregated state signalled by persistent "loyalty" of certain groups of traders to certain markets.

A. Alorić (✉) • P. Sollich
Disordered Systems Group, Department of Mathematics, King's College London, Strand, London WC2R 2LS, UK
e-mail: aleksandra.aloric@kcl.ac.uk

P. McBurney
Department of Informatics, King's College London, Strand, London WC2R 2LS, UK

F. Amblard et al. (eds.), *Advances in Artificial Economics*, Lecture Notes in Economics and Mathematical Systems 676, DOI 10.1007/978-3-319-09578-3_7

In order to test our hypothesis that segregation can emerge spontaneously, we have constructed a simple model of markets and traders. Markets are governed here by simple static sets of rules—how to set the trading price and how to match traders. Traders are taken as Zero Intelligence agents following Gode and Sunder (1993). Such traders act largely randomly. This makes them a convenient tool for investigating the impact of market mechanisms (Ladley 2012), by removing all of the complexity associated with the traders' strategies. However, we note that our agents are zero intelligence only with respect to price, i.e., they generate bids and asks at random. On the other hand they do learn from past successes or failures about the choice of market and whether to buy or sell.

Closely related work on segregation in Hanaki et al. (2011) studies agents competing for parking spots in a one way street. A learning process is again used, with rewards (the closer to the city centre the better) and penalties (an agent is punished if she/he reaches the city centre without parking). It is shown that the population splits into two groups, agents who persistently choose parking spots close to the city centre on the one hand and agents settling for spots further away on the other. Grouping of agents in an economic context was studied in multi-resource minority games (Huang et al. 2012). In this model, grouping emerges when the probability that an agent will copy the strategy of a winning neighbour is large enough. However, in contrast to our model, it was assumed in this scenario that there is a considerable amount of structure in the connectivity among traders, as well as perfect information about the actions of neighbours.

2 Model

We consider a simplified model of markets and decision-making traders with the aim of investigating the segregation of traders. During each trading period agents are confronted with a choice of actions: where to trade—*choice of market*—and how to trade—*whether to act as buyer or seller.* Decisions are made based on the attractions, which are accumulated scores an agent has received when taking actions in the past. The attractions to the various actions are updated after every trading period using a reinforcement learning rule of the form[1]

$$A_\gamma(n+1) = \begin{cases} (1-r)A_\gamma(n) + rS_\gamma(n), & \text{if agent has chosen action } \gamma \\ (1-r)A_\gamma(n), & \text{if agent has chosen action } \beta \neq \gamma \end{cases}$$

[1]Hanaki et al. (2011) uses the same rule with $\omega = 1 - r$, while in Sato and Crutchfield (2003) the prescription used was $A(n+1) = S(n) + (1-\alpha)A(n)$. The second rule allows the attractions to increase to infinity, while in the first case, they are constrained. However, up to a temperature rescaling, the two rules are equivalent. The more important difference is that in the paper (Sato and Crutchfield 2003), the attractions of unplayed actions are updated with fictitious scores an agent would have got had he played the action, while we effectively update them with score $S(n) = 0$.

where $S_\gamma(n)$ is the return gained by taking action γ for the nth trade; r is the parameter that describes the agent's memory. Its intuitive meaning is that each attraction is effectively an average of the returns over a shifting time window covering the previous $\frac{1}{r}$ trades. Finally, $A_\gamma(n+1)$ is attraction to the action γ after n trades, which will determine the action chosen in the following $(n+1)$th trading period. The choice of action is then calculated using the *softmax*[2] function: the probability of taking an action γ is $P_\gamma \propto \exp(A_\gamma/T)$. The temperature T regulates how strongly agents bias their preferences towards the option that gathered them the highest score. For $T \rightarrow 0$ agents strictly choose the option with the highest attraction, while for $T \rightarrow \infty$ they choose randomly among the options.

Orders to buy/sell at a certain price (bids and asks) are generated by traders independently of previous success or any other information; the bids and asks are independently identically distributed random variables (thus Zero Intelligence). We assume bids (b) and asks (a) are normally distributed ($a \sim \mathcal{N}(\mu_a, \sigma_a^2)$ and $b \sim \mathcal{N}(\mu_b, \sigma_b^2)$), with means satisfying $\mu_b > \mu_a$. The assumption that the average bid is higher than the average ask is not crucial; it mainly allows a larger number of successful trades as the resulting trading price is typically below the average bid and above the average ask. In the work of Gode and Sunder (1993) various demand and supply curves were used and thus both orderings of average bids and asks, $\langle a \rangle > \langle b \rangle$ and $\langle a \rangle < \langle b \rangle$, were investigated: they lead qualitatively to the same results. We similarly explored the case $\mu_a > \mu_b$, and apart from the obvious quantitative consequence that a smaller fraction of orders is valid for trade and consequently the number of successful trades is smaller, the qualitative results remain the same. Once all traders have submitted an order to the market of their choice, then at each market the average bid $\langle b \rangle$ and average ask $\langle a \rangle$ are calculated and the trading price is set as $\pi = \langle a \rangle + \theta(\langle b \rangle - \langle a \rangle)$ with θ being a parameter that describes the bias of the market towards buyers (for $\theta < 1/2$) or sellers (for $\theta > 1/2$).[3] All buyers who bid less and all sellers who ask more than the trading price are removed from the trading pool, as their orders cannot be satisfied at the price that has been set. The remaining traders are matched in random pairs of buyers and sellers, giving a total number of trades $\min(N_{\text{valid bids}}, N_{\text{valid asks}})$. For traders who manage to trade, the score is calculated as:

$S(n) = \pi - a_n$ (sellers value getting more than they were asking for, i.e. a_n)
$S(n) = b_n - \pi$ (buyers value when they pay less then they intended, i.e. b_n)

[2]The softmax function is commonly used in models of learning agents, see for example Hanaki et al. (2011), Sato and Crutchfield (2003). Another common formulation of the softmax function is $P_\gamma \propto \exp(\beta A_\gamma)$, where $\beta = 1/T$ is sometimes called *the intensity of choice* as in Brock and Hommes (1997).

[3]Note that traders are not informed about these market biases, nor the market mechanism in general; they learn only by means of the scores they receive.

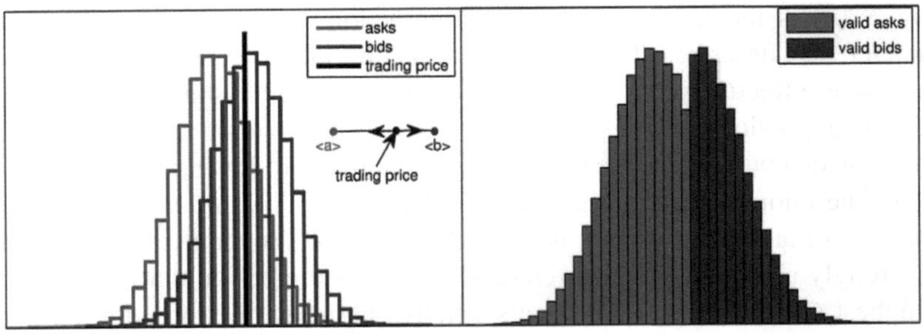

Fig. 1 Illustration of market mechanism. (*Left*) Histogram of bids and asks arriving at a given market. The *inset* shows how the trading price is set, with a bias towards average bid or average ask regulated by the bias parameter θ of the market. (*Right*) Once invalid orders are eliminated, i.e. bids below or asks above the trading price, the distributions of valid bids and asks remain. Traders who have submitted valid orders are matched in random buyer-seller pairs for trading

All traders who do not get to trade receive return $S(n) = 0$, and all orders are deleted from the market after each trading period. Figure 1 illustrates this market mechanism.

The assignment of returns that we are using was introduced in Gode and Sunder (1993), where it is associated with budget constraints of "Zero Intelligence-Constrained" traders. Exactly these agents were shown to reproduce the efficiency of human traders in double auction markets. In the original work, a_n are cost values assigned to sellers, while b_n are redemption values assigned to buyers. Traders were allowed to trade only if the trading price was lower than the redemption value or higher than the cost value, thus the name *constrained* agents. Although the assignment of returns is the same in our model, we do not use the term budget constrained in the description as our agents are allowed to persistently buy (or sell), which is possible only if there is no overall wealth constraint.[4] In our model the bids and asks could similarly be interpreted as cost and redemption values. We assume in addition that agents determine their orders based on these values, while the actual trading price is a function of the population averages.

3 Results and Discussion

In this section we will present the results from the simulations of the trading system described so far in this paper. The system parameters are the number of agents N, the number of markets $M = 2$, the biases of the markets θ_1, θ_2, the means

[4]We note that also in Gode and Sunder (1993), agents were preassigned the role of a buyer or a seller and were not allowed to change this during trading, thus acting as if there was no overall constraint on the possession of money/goods for trade.

and standard deviations of the distributions of bids and asks $\mu_a, \sigma_a, \mu_b, \sigma_b$, the temperature T and the forgetting rate r. For every set of parameters simulations were run for 10,000 trading periods; statistics are presented for data gathered from the last 100 trading periods of 100 independent runs of the stochastic dynamics.

In our system each agents has four preferences $p_{\mathcal{B}1}$, $p_{\mathcal{B}2}$, $p_{\mathcal{S}1}$, $p_{\mathcal{S}2}$ for the four possible actions of buying and selling at market 1 or 2. In the figures below, to help visualization we represent each agent by their total preference for buying ($p_{\mathcal{B}} = p_{\mathcal{B}1} + p_{\mathcal{B}2}$) and for market 1 ($p_1 = p_{\mathcal{B}1} + p_{\mathcal{S}1}$). This is convenient as the corners in the $(p_{\mathcal{B}}, p_1)$ plane then represent the four *pure strategies*—agents *always* buying at market 1, etc. Similarly, in the space of attractions we use two coordinates $(\Delta_{\mathcal{BS}}, \Delta_{12})$, which are basically attraction to buying as against selling and attraction to market 1 as against market 2.

In Fig. 2 we present steady state attraction and preference distributions for temperature $T = 0.29$. An initially narrow, *delta peaked* distribution (all scores are equal to 0) has been broadened due to diffusion arising from the random nature of returns. This steady state represents unsegregated behaviour of a population of traders. While the population does include some traders with moderately strong preferences for one of the actions, preferences remain weak on average. The population as a whole remains homogeneous in the sense that there is no split into discernible groups.

Figure 3 contrasts this scenario with the steady state of a system with exactly the same set of parameters but at the lower temperature $T = 0.14$. The population of traders now splits into four groups, with the agents persistently trading at one of the markets, and thus we call this state *segregated*. The markets shown in this example (Figs. 2 and 3) are biased so that if an agent buys at market 1, or sells at market 2 (actions $\mathcal{B}1$ or $\mathcal{S}2$) he is awarded with a higher score. The traders who prefer these actions are *"return-oriented traders"*. However, if all traders were return-oriented, they would have no partners for trading, and consequently they would receive zero scores. We see that to enable trading, some traders develop strong

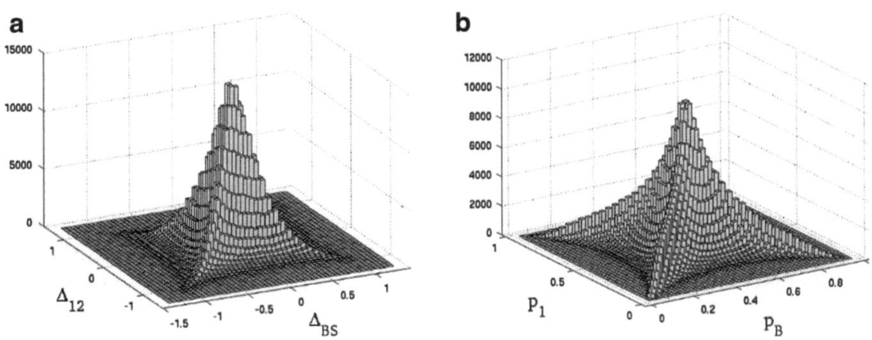

Fig. 2 Steady state distributions at temperature $T = 0.29$, with other parameters set to $N = 200$, $M = 2$, $\theta_1 = 0.3$, $\theta_2 = 0.7$, $r = 0.1$, $\mu_b - \mu_a = 1$, $\sigma_a = 1$, $\sigma_b = 1$. (**a**) Distribution of attractions. (**b**) Distribution of preferences

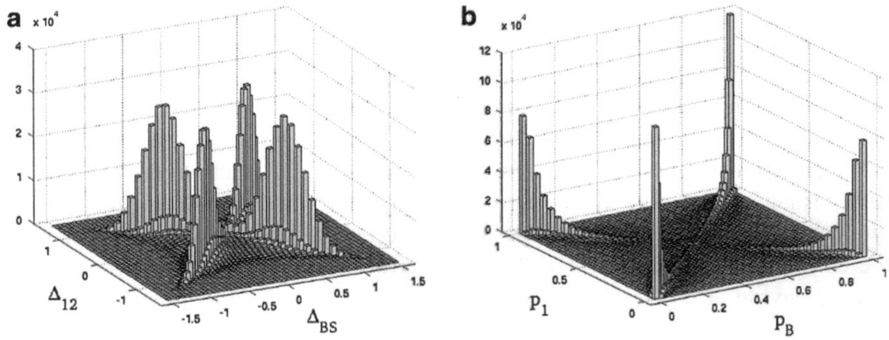

Fig. 3 Steady state distributions in the low temperature regime ($T = 0.14$, all other parameters as in Fig. 2), showing clearly the segregation of traders into groups. (**a**) Distribution of attractions. (**b**) Distribution of preferences

preferences for buying (selling) at a market that gives them a lower average return ($\mathcal{B}2, \mathcal{S}1$). A larger fraction of these traders will be removed from the market as their orders will be regarded as invalid more frequently. Consequently, these traders will form a minority group and they will always find a trading partner, hence we will call them *"volume-oriented traders"*. The occurrence of segregation of an initially homogeneous population of traders into groups of return-oriented and volume-oriented traders is the main qualitative result of this paper.

When assessing stationarity of our system we measured population and time averages for various observables ($A_\gamma, \Delta_{\mathcal{BS}} \ldots$). Depending on parameters, a stationary state was generally reached reasonably quickly, mostly within $1,000$ trading periods. Apart from stationarity we also investigated to what extent our system is ergodic, i.e. we wanted to exclude possibility that distributions in the low temperature regime might be a consequence of some agents' preferences becoming essentially *frozen* after the first few trades. Quantitatively, we measured persistence times in one of four quadrants—"prefer buying at market 1" ($\Delta_{\mathcal{BS}} > 0$ and $\Delta_{12} > 0$), "prefer selling at market 1" ($\Delta_{\mathcal{BS}} < 0$ and $\Delta_{12} > 0$), etc. Figure 4 shows the average time an agent spent in any one of these quadrant before leaving it for another quadrant, for various temperatures. We present these plots for different values of the forgetting rate r, using the rescaled time $t = rn$ where n is the number of trading periods. (The use of t rather than n ensures that the trivial effect on persistence times of agents updating their attractions more slowly at smaller r is removed.) From the figure one sees that at small enough r, the onset of segregation is accompanied by a rapid increase in persistence times, showing that in the segregated state agents do indeed remain "loyal" to a given market for long times. On the other hand, we see that when temperatures are not too low (i.e. above the levelling off of the small-r curves in Fig. 4) then persistence times are short compared to the overall length of our runs, so that the system is ergodic.

To quantify the observed change in the distributions of agent attractions or preferences as we go from unsegregated to segregated states, we measured higher

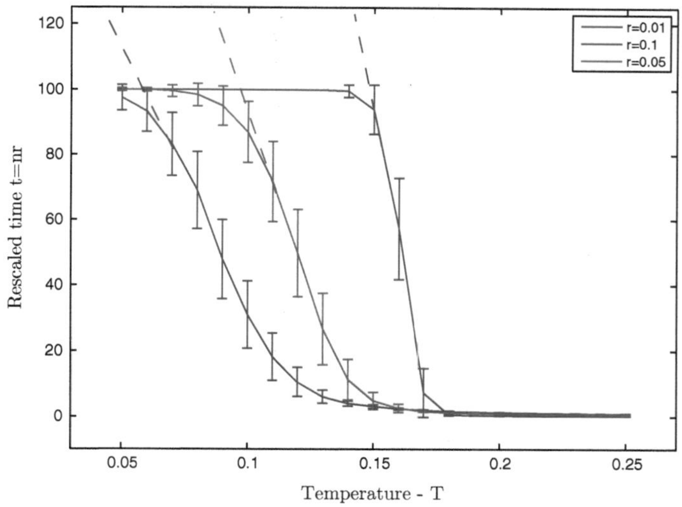

Fig. 4 Average time an agent persists in any one of the four preference quadrants, plotted against temperature for different values of the forgetting rate, $r = 0.1$ (*blue*), $r = 0.05$ (*red*) and $r = 0.01$ (*green*). *Dashed lines* are sketches of how the persistence times would increase further if they were not limited by the length of our simulation runs. Data were taken from longer periods to enable in each case scaled persistence times up to $t = 100$ to be measured. Other parameters (as previously): $N = 200$, $M = 2$, $\theta_1 = 0.3$, $\theta_2 = 0.7$, $\mu_b - \mu_a = 1$, $\sigma_a = 1$, $\sigma_b = 1$ (Color figure online)

cumulants of the distributions $P(\Delta_{BS})$ and $P(\Delta_{12})$. Specially we tracked the *Binder cumulant*: $B = 1 - \frac{\langle \Delta^4 \rangle}{3\langle \Delta^2 \rangle^2}$. Figure 5 shows values of this Binder cumulant for various temperatures of the system, with all other parameters being same as in the previous figures. For higher temperatures, the Binder cumulant of our distributions approaches value characteristic of Gaussian distributions ($B = 0$) as expected. At the other extreme, in the low temperature regime, the cumulant approaches a second characteristic value $B = 2/3$, which is the Binder cumulant of a distribution consisting of two sharp peaks with equal weight. The transition between these two regimes is sharper for smaller values of r, making it possible to estimate the critical temperature for the onset of segregation.

Our simulation results suggest that even our simplified trading system shows rich and interesting behaviour. There exists a critical temperature T_c, such that for values $T < T_c$ the system segregates, i.e. the population of initially homogeneous traders splits into groups that persistently choose to trade at a specific market. The persistence times increase strongly with decreasing forgetting rate r (see Fig. 4) and we conjecture that in the limit $r \to 0$ there is a sharp transition at T_c in the sense that the persistence time diverges there. The exact value of the critical temperature is a function of the market parameters, and for the values of $\theta_{1,2}$ used above, we would estimate it from Fig. 5 to be $T_c \approx 0.17$.

To understand in more detail how segregation arises, and the nature of the transition to the segregated state as T is lowered, a simple mathematical description

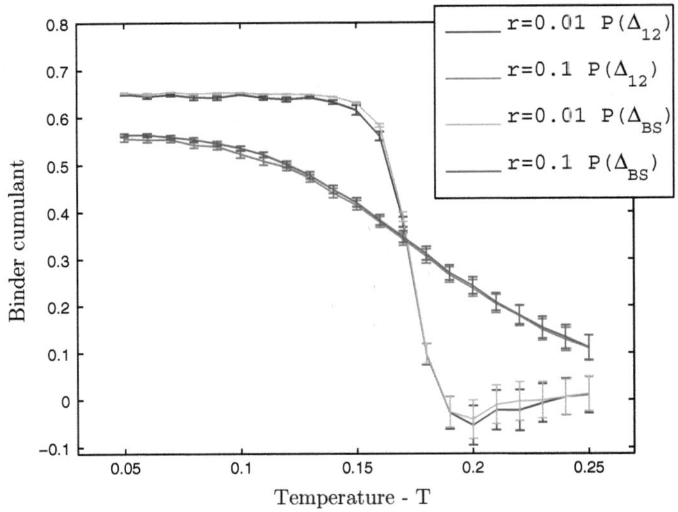

Fig. 5 Binder cumulant for $P(\Delta_{BS})$ and $P(\Delta_{12})$ distributions versus temperature for two different values of the forgetting rate, $r = 0.1$ and $r = 0.01$. Other parameters as previously $N = 200$, $M = 2$, $\theta_1 = 0.3$, $\theta_2 = 0.7$, $\mu_b - \mu_a = 1$, $\sigma_a = 1$, $\sigma_b = 1$

would evidently be useful. To obtain such a description, we can build on the approach of Brock and Hommes (1997). This work studies the dynamics of agents who have to decide whether to purchase a sophisticated price predictor, or use a freely available naive predictor of price. This scenario differs from our model in a number of ways; apart from the more sophisticated trading strategies of the agents, it assumes perfect information about previous prices and about the performance of any price predictor. What is important in the analysis of Brock and Hommes (1997), however, is that the limit of a large population of agents is implicitly taken, so that the system can be described entirely in terms of the fraction of agents choosing a given action (price predictor) at any instant in time, with these fractions evolving deterministically in time. The authors of Brock and Hommes (1997) show that depending on the temperature, or the "intensity of choice" $\beta = 1/T$, these two fractions can exhibit rich dynamics. The origin of this is that when all traders use sophisticated predictors, the cost of this predictor leads some agents to start choosing the free predictors, while there is a reverse effect from positive feedback when all traders use the simple predictor.

To adopt a similar approach for our model, we realize that mathematically our dynamics is Markovian, provided that we keep track of the attractions A^i_γ to all actions $\gamma = \mathcal{B}1, \mathcal{S}1, \mathcal{B}2, \mathcal{S}2$ of all agents $i = 1, \dots, N$. Working with this description in a $4N$-dimensional continuous state space is, however, very difficult. As in Brock and Hommes (1997) we can therefore consider the large N-limit where the trading price at each market is no longer affected by fluctuations in the number and value of orders submitted. We also consider the limit of small r, using as time unit again the rescaled time $t = rn$ so that a unit time interval in

t corresponds to $1/r$ trading periods. The fluctuations in each individual agent's attractions then also tend to zero because they are averaged over many ($\sim 1/r$) returns each contributing a small ($\sim r$) change of attraction. As long as the agent population remains homogeneous, all agents should in the limit have the same attractions A_γ. In that case the system is described entirely in terms of the average values of these four attractions, or correspondingly the fraction of agents choosing each of the four options γ. As these fractions add to unity it is enough to keep track of three of them, and one can write down deterministic equations for their time evolution. (Details are beyond the scope of this paper and will be given elsewhere.)

The results of the above approach for our model are still somewhat difficult to visualize as we need to track fixed points and trajectories in a three-dimensional space. We therefore switch to a simpler system that gives qualitatively similar results: a population of traders consisting of two equal-sized groups with fixed preference for buying $p_{\mathcal{B}}^{(1)}$ and $p_{\mathcal{B}}^{(2)}$, respectively. The agents then only choose between two actions, namely, whether to go to market 1 or 2 in each trading period. Although the system where agents change their buy-sell preferences is more plausible behaviourally, the two-group model still undergoes segregation and requires us (for $N \to \infty$, $r \to 0$ and assuming an unsegregated state as above) to track only the fraction of agents choosing market 1 in each of the two groups. We denote these fractions by $f^{(1)}$ and $f^{(2)}$. In Fig. 6 we present the flow diagrams that we find for the time evolution of these two fractions, at high and low T. At high temperature, one observes a single fixed point as expected (Fig. 6a). As T is lowered, this fixed point becomes unstable, and two additional stable fixed points appear

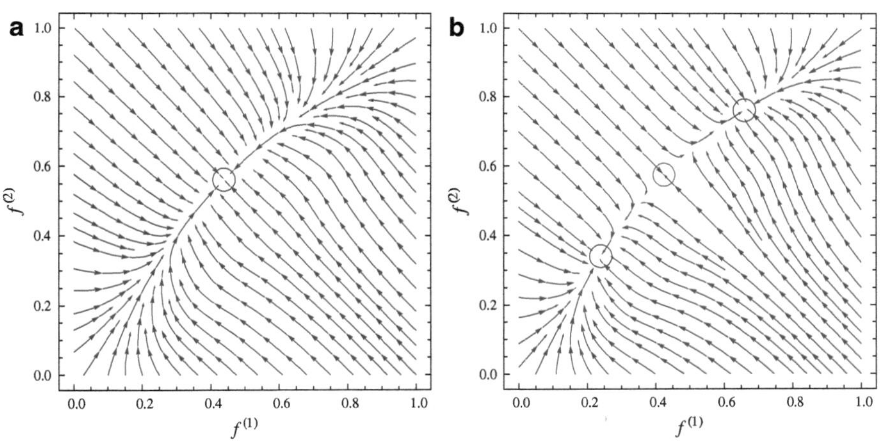

Fig. 6 Flow diagrams that describe the large population dynamics of our two-group model in the space of fractions of agents from each group who choose market 1. $f^{(1)}$ is the fraction going to market 1 in the group of agents who typically sell ($p_{\mathcal{B}}^{(1)} = 0.2$), and $f^{(2)}$ the corresponding fraction in the group of "buyers" ($p_{\mathcal{B}}^{(2)} = 0.8$). The markets have symmetric biases $\theta_1 = 0.3 = 1 - \theta_2$. (**a**) High temperature ($T = 0.32$): the dynamics has a single fixed point. (**b**) Low temperature ($T = 0.29$): the single fixed point has become unstable

(Fig. 6b). The temperature where the high-T fixed point first becomes unstable thus identifies the critical temperature T_c for the onset of segregation. We also find that the new stable fixed points evolve continuously from the high-T fixed point as T is lowered through T_c, so the segregation transition has the character of a bifurcation and is continuous.

It is worth emphasizing that the locations of the new fixed points that appear at low temperature are not necessarily meaningful: as explained above, the simplifications that have allowed us to consider deterministic time evolution in a simple two-dimensional space require that the agent population remains homogeneous. By construction, this simple picture can therefore not describe quantitatively the segregated populations of agents that arise below T_c. Nevertheless, the instability of the high-T unsegregated fixed point is enough to identify the temperature for the onset of segregation.

The analytical description sketched briefly above allows us to study, for example, how the value of the critical temperature T_c depends on the parameters of the problem, specifically for the two-group model on $p_{\mathcal{B}}^{(1)}$ and $p_{\mathcal{B}}^{(2)}$ and on the market biases θ_1 and θ_2. As an example, Fig. 7 shows how T_c varies with the market bias, still for the case of symmetric markets $\theta_1 = 1 - \theta_2 = \theta$. One sees that for every value θ there exists a critical temperature T_c at which a bifurcation to a segregated steady state occurs. Note that the temperature region where segregation occurs *shrinks* as the difference between the market biases *increases* (smaller θ), showing that segregation is a collective effect rather than being trivially driven by the differences between the markets. For $\theta = 0.3$ as in Fig. 6, one finds $T_c(\theta) = 0.308$. From simulations for a system with $N = 100$ traders and forgetting rate $r = 0.1$,

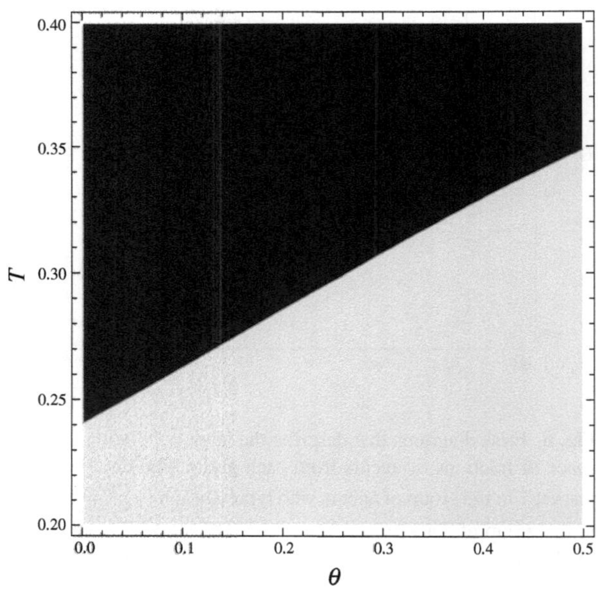

Fig. 7 Segregation temperature T_c versus market bias $\theta_1 = 1 - \theta_2 = \theta$. In this diagram $T_c(\theta)$ separates segregated (*beige*) and unsegregated (*dark blue*) steady states. Results are shown for the two-group model with the two groups of agents having fixed buying preferences of $p_{\mathcal{B}}^{(1)} = 0.2$ and $p_{\mathcal{B}}^{(2)} = 0.8$, respectively (Color figure online)

we estimate a value of $T_c \approx 0.3$. This constitutes excellent agreement with the theoretical prediction, especially bearing in mind that the latter applies directly only to the limit $N \to \infty$ and $r \to 0$.

In our original model where the agents can adapt their preferences both for the two markets and for whether to buy or sell, the quantitative agreement is slightly less good. For example $\theta_1 = 1 - \theta_2 = 0.3$ and bid and ask distribution parameters as in Figs. 2 and 3 the analytical description predicts $T_c \approx 0.157$. Our simulations for a population of $N = 200$ traders with forgetting rate $r = 0.1$, on the other hand, lead to the estimate $T_c \approx 0.17$. This suggests that in the fully adaptive model the effects of nonzero forgetting rate and finite population size are stronger than in the two-groups model.

Concluding Remarks

With so much trade and commerce moving online over the last two decades, the study, design, operation, and good governance of electronic marketplaces has become a major area of computer science, both theoretical and applied. Much online economic activity—for example, most trading in western financial markets—is now undertaken by automated computer programs, which are software agents acting on behalf of human principals or companies. A key research goal in the study of electronic marketplaces is, therefore, to understand the long-run dynamics of these markets when populated by automated software traders. This leads to questions such as: what long-run states are possible in these marketplaces, what patterns in states occur or recur, what states may be avoided and how, what states may be encouraged to occur and how, etc. The practical economic and financial consequences of such understanding are immense. The so-called *Flash Crash* of US stock markets on 6 May 2010 showed the vulnerability of inter-linked trading systems to a single large trade, for example, and has led to the implementation of automated circuit breakers to eliminate or reduce the sector-wide impacts of rapid market movements (Findings Regarding the Market Events of May 6 2010). The importance of these issues is shown by the establishment of a major research programme by the UK Governments Department of Business, Industry and Skills on computer trading in financial markets.[5] Our research in this same vein focuses on a description of a specific characteristic of trading systems—segregation. As argued in the introduction, specialized (segregated) traders might be better in terms of exploitation of a market. However with specialization there comes an associated vulnerability as agents become more exposed to losses if all their investments are focused on a single market that might crash. Ultimately, we would like to describe and predict the long-run dynamics of marketplaces

(continued)

[5] See: http://www.bis.gov.uk/foresight/our-work/projects/published-projects/computer-trading.

comprising automated interacting traders and to extract a set of regulations that might promote or suppress the segregation.

In this paper we introduced a simplified model of double auction mechanisms with Zero Intelligence traders, with the goal of investigating the possibility of a spontaneous segregation of traders. The use of ZI traders was motivated by the hypothesis that segregation can emerge as a consequence of market mechanisms and learning rules, neglecting complexity in trading strategies. We presented results form numerical simulations and outlined how analytical methods can be used to understand the occurrence of segregation, giving quantitatively reasonable predictions even away from the limits (infinite population of traders, infinitesimal forgetting rate) where the analysis is derived. Although the relevance of our model with respect to real economies might be questioned due to its simplicity, it is interesting to note that even with this simple trading mechanism, learning agents who interact only via markets can end up being segregated.

References

Brock WA, Hommes CH (1997) Rational route to randomness. Econometrica 65(5):321–354

Cai K, Gerding E, McBurney P, Niu J, Parsons S, Phelps S (2009) Cat overview. Technical report, University of Liverpool

Findings Regarding the Market Events of May 6, 2010 (2010) Report of the staffs of the CFTC and SEC to the joint advisory committee on emerging regulatory issues

Gode DK, Sunder S (1993) Allocative efficiency of markets with zero-intelligence traders: market as a partial substitute for individual rationality. J Polit Econ 101(1):119–137

Hanaki N, Kirman A, Marsili M (2011) Born under a lucky star?. J Econ Behav Organ 77(3):382–392

Huang Z, Zhang J, Dong J, Huang L, Lai Y (2012) Emergence of grouping in multi-resource minority game dynamics. Sci Rep 2:703

Ladley D (2012) Zero intelligence in economics and finance. Knowl Eng Rev 27(2):273–286

Nagarajan G, Meyer RL, Hushak LJ (1995) Segmentation in the informal credit markets - the case of the Philippines. Agric Econ 12(2):171–181

Robinson E, McBurney P, Yao X (2012) Co-learning segmentation in marketplaces. In: Adaptive and Learning Agents, vol 7113. Springer, Berlin, pp 1–20

Sato Y, Crutchfield JP (2003) Coupled replicator equations for the dynamics of learning in multiagent systems. Phys Rev E 67(1):015206

The J-Curve and Transaction Taxes: Insights from an Artificial Stock Market

Lina Kalimullina and Rainer Schöbel

1 Introduction

A significant underperformance of the majority of actively managed funds in comparison with the market average raises the question whether additional information can be beneficial to agents. Studies with informed agents and non-informed agents agree that knowing more is better than knowing less (Grossman and Stiglitz 1980). Following the argumentation in Fama (1970), only in efficient financial markets does obtaining additional information not result in increased returns, since all information is already reflected in market prices. However, "strong" market efficiency is hardly possible in reality. Therefore information may be beneficial to market participants. According to Schredelseker (1984), economic intelligence is believed to have an increasing function: the more information one has, the better results one can obtain. Nevertheless, when there are several information levels present, a more complex relation can be observed (Huber 2007; Tóth and Scalas 2008). Various studies of experimental financial markets have discovered J-shaped relative returns among information levels: the best informed agents outperform the market, but only averagely informed agents perform worse than the least informed or completely uninformed agents (Huber 2007). These results suggest that partial information may be harmful.

In this paper, we examine the J-curve existence in a simulated financial stock market. In addition, we explore a case in which the RR's function has a different shape.

L. Kalimullina • R. Schöbel (✉)
Faculty of Economics and Social Sciences, University of Tübingen, Mohlstrasse 36, 72074 Tübingen, Germany
e-mail: lina.kalimullina@uni-tuebingen.de; rainer.schoebel@uni-tuebingen.de

F. Amblard et al. (eds.), *Advances in Artificial Economics*, Lecture Notes in Economics and Mathematical Systems 676, DOI 10.1007/978-3-319-09578-3_8

The discussion concerning the pros and cons of implementing transaction taxes in financial markets is still viable. Supporters claim that taxes reduce market volatility. Opponents argue that taxes result in a decreased market efficiency. Apart from checking the effects of a tax levy on these market parameters, we also investigate whether this levy causes any changes to the J-curve.

In order to achieve our goals, we reproduce from scratch an agent-based ASM, as described in Tóth et al. (2007), Tóth and Scalas (2008); it models heterogeneously behaving agents who possess heterogeneous levels of information. This approach allows us to incorporate and replicate stylized properties of real financial markets (Westerhoff 2010) as well as market imperfections (Levy et al. 2000).

The rest of the paper is organized as follows. In Sect. 2, we briefly describe the ASM implemented in Matlab. In Sect. 3, we discuss the J-structure of final returns among the varying information levels resulting from extensive simulations, and explore a case when this shape is not valid. Section 4 extends the discussion by introducing transaction taxes and examines how this affects the market parameters and the J-curve. Section "Conclusion" offers a conclusion.

2 Market Structure

In this section, we describe the continuous double auction (CDA) trading system, the economic environment and the market process sequence. We focus on two sources of agent's heterogeneity: the amount of information they possess ("informational endowment") and their trading strategies.

2.1 Trading System

The marketplace mechanism, where the selling and buying of stocks occurs, is modeled as an order-driven CDA. This trading mechanism is common to many stock exchanges (Anufriev et al. 2013). In such a market, agents trade directly with each other and the market maker does not participate in transactions. As in Cervone et al. (2009), traders arrive at the market sequentially and place either marketable orders, leading immediately to a transaction, or limit orders, which are written down in the order book. By allowing traders to send limit orders, liquidity is secured. If an agent wants to buy the stock, he submits an order to buy–a bid order; if he wants to sell the stock, he submits an order to sell–an ask order. Trade is facilitated by the order book that represents a queue of limit orders which, once sent to the market, stay there until they are satisfied or until the end of a period of time. In the limit order book, the price is set in multiple steps when a newly placed market order matches any limit order previously sent in by another agent. As in Anufriev et al. (2013), the best limit order serves as the transaction price if orders are matched. Furthermore, transactions follow a time priority rule for orders with identical prices, and a price priority rule for orders across all price levels (Chiarella and Iori 2002).

2.2 Economic Environment

The economic environment of our ASM is similar to those presented in Huber (2007), Tóth and Scalas (2008). We designed a multi-period market where heterogeneously informed artificial agents (computer algorithms) trade a risky asset, and possess a portfolio consisting of this risky asset (stock) and a risk-free asset (cash). In our simplified case, only one risky asset is available for trading. Going short in cash or stock is not allowed. At the end of each period, interest on the cash account and dividends on stocks are paid out. Dividends are simulated as a random walk process before each trading session:

$$D_t = |D_{t-1} + 0.1 \times N(0,1)| \;\; \forall t = 1 \ldots T,$$

where T is the number of simulated periods and $N(0,1)$ denotes a standard normally distributed random variable. The starting value of dividend is $D_0 = 0.2$. The dividend process is assumed to be non-negative, meaning that whenever $D_t < 0$, the absolute value is taken for the purpose of the dividend process (Tóth et al. 2007; Tóth and Scalas 2008).

As in Schredelseker (1984), we define our market as a pure circulation market: no new stocks are issued. The market is populated with ten heterogeneous agents; each of them possibly has a different information endowment and uses a different trading strategy. Information about orders and past prices is public; it is available to all agents free of charge at all times. The final endowments of stock and cash resulting from one period are carried over to the next period.

2.3 Information Structure

Each trader is endowed with some information about future dividends depending on his information level. The mechanism of information allocation takes the form of a moving window: the best informed agent has the most comprehensive information, the second best informed trader obtains the same information set but after one period, while the least informed agent has access to the same information set after the time lag of eight periods. At the beginning of each new period, an agent receives additional information about one further dividend realization.

Agent I_0 (information level of zero) does not possess any information. Such an agent is also called a completely uninformed agent; he trades randomly on the market. Agent I_1 has information about dividend payments at the end of the current period; he is called the least informed agent. In general, an agent I_j receives information about j future dividends. The best informed agent is I_9. The better informed agents have an informational advantage since they gain access to desirable information earlier than the worse informed agents (Huber 2007). It is assumed that information is exact, and dividend values do not suffer from any noise or inaccuracy.

Once the information about future dividends is obtained by traders, it is modified in the subsequent prediction process. We assume that all agents use the same prediction mechanism. With the help of Gordon's formula, agents calculate the expected present value of stock (EPV) conditioned on their forecasting horizon (Tóth and Scalas 2008):

$$EPV_{j,k} = \frac{D_{k+j-1}}{r_e(1+r_e)^{j-2}} + \sum_{i=k}^{k+j-2} \frac{D_i}{(1+r_e)^{i-k}}.$$

Here $EPV_{j,k}$ denotes the conditional EPV of a stock for a trader I_j with information level j in a period k, and r_e is a risk-adjusted interest rate.

In contrast to the experimental market with human agents in Huber (2007), in our simulation set-up there is no feedback on the information level: an agent does not realize that the other agents might be better informed than himself.

2.4 *Trading Strategies of Market Participants*

Artificial agents use certain algorithms to decide on a market transaction. Three types of agents are modeled to interact in the market. We model random traders, fundamental traders, and chartists, as described in Tóth and Scalas (2008).

A random trader constructs a new limit order as a random deviation from the previous transaction price. We assume that random traders believe that they are trading based on information which they think they possess; or that they simply like to trade (Black 1986).

The main assumption about fundamentalists is their belief that information about future changes of fundamental value is not reflected in the price process yet but that the price will converge to its fundamental value in the future (Schredelseker 1980). Their trading decisions are based on EPVs. If the EPV is larger (smaller) than the best ask (bid), it is treated as a new marketable bid (ask) order.

A chartist (or technical trader) ignores the information about the future dividends and follows a trend (or momentum) strategy. He analyzes only price changes and does not conduct any economic analysis. His decisions are based on the presence of uptrends or downtrends or on the absence of any type of clear trends in the price evolution process. Including a technical trader in the simulated market is a natural choice, in light of the conclusions from the experimental results in Levy et al. (2000): human agents, even after being told that the RRs are randomly drawn variables, tend to attach considerable importance to past performance; the majority of traders believe that the future will be either like the past or contrary to the past.

It is assumed that agents cannot change their strategy, they lack social interaction and the ability to evaluate success or failure. Learning or adaptive mechanisms are not incorporated.[1]

2.5 Market Session Sequence

The following steps summarize the sequence of main activities of the simulated ASM. After the dividend process is randomly simulated and the agents are assigned exogenously with information levels, each of them receives a corresponding amount of information about future dividends. The market session begins with the "opening" procedure, which sets the starting price through a Walrasian auction by determining the equilibrium price of supply and demand (Bauwens and Giot 2001). In the following market session, agents are chosen randomly to act sequentially in the market. If a transaction is executed, the wealth bookkeeping mechanism changes the current endowments of the agents. At the end of each period, the order book is emptied: all the unmatched limit orders are removed from the order book. Moreover, dividend and interest payments are added to the final endowments.

2.6 Simulation Set-Ups

We conducted two extensive simulations with slightly changed parameters.

The first set of parameters corresponds to the simulation for Sect. 3. The market is populated with ten agents; only one of them is a random trader while the others are fundamentalists. There is only one fundamental trader per information level. Chartists are not included in the market population at this stage as they do not use the fundamental information (later on we describe how including the chartists influences the simulation results). The market session lasts 100 periods, each consisting of 100 steps. Initially the agents are endowed with high amounts of cash and stock, so that they are very unlikely to face any budget constraints during a market session: the initial cash endowment is 10,000 units of currency and the initial stock endowment is 1,000 units. The initial price of the stock is 40 units of currency. For the purpose of statistical reliability, we conduct 100,000 market iterations (100×100 simulation): 100 sessions are completed with different random dividend processes and 100 further runs are repeated within each session under the same dividend process to account for randomness in the sequence of agents' draw and market actions.

The second set of parameters corresponds to the simulation for Sect. 4. In comparison to the previous set-up, the market session lasts 75 periods. The initial

[1]Learning mechanisms in a CDA are considered in Posada et al. (2006a,b).

cash endowment is 1,000 units of currency and the initial stock endowment is 100 units. The simulation repeats 2,800 market iterations for each unique set of changing market parameters (70×40 simulation): there are 70 sessions with different random dividend processes and 40 further runs within each session.

3 Evaluation of the Agents' Success: The J-Curve

The J-curve of RRs, thoroughly investigated in Schredelseker (1984) for the first time, describes the inequality of RRs obtained by the heterogeneously informed agents. In this section, we investigate the J-curve resulting from the simulation of our artificial market model.

We assess the final wealth allocation with the overall relative return (ORR). ORR is the RR estimated from the market opening until its closing, minus the market return. The market return is calculated as the average individual return and is taken as a benchmark for the comparison of returns among the agents.

The Mann-Whitney U-test confirmed the J-shape for our market setting: not all informed traders perform better than the random trader, see Table 1 and Fig. 1.

The ORRs of the uninformed I_0 and medium informed I_6 agents are very close to zero. The random trader ends up with the market average return, which is expected

Table 1 Mann-Whitney U-test for pairs of ORRs

Agent	I_0	I_1	I_2	I_3	I_4	I_5	I_6	I_7	I_8	I_9
I_0	–	1	1	1	1	1	0	1	1	1
I_1	1	–	0	0	0	1	1	1	1	1
I_2	1	0	–	0	0	1	1	1	1	1
I_3	1	0	0	–	0	1	1	1	1	1
I_4	1	0	0	0	–	1	1	1	1	1
I_5	1	1	1	1	1	–	1	1	1	1
I_6	0	1	1	1	1	1	–	1	1	1
I_7	1	1	1	1	1	1	1	–	1	1
I_8	1	1	1	1	1	1	1	1	–	1
I_9	1	1	1	1	1	1	1	1	1	–

Note: The mean ORRs are compared pairwise (I_0 with I_1, I_1 with I_2, and so forth). The null hypothesis H_0 states that the two samples are drawn from the distributions with equal medians or, in other words, that two information levels have equal average ORRs. In the table the logical value h is presented: it represents the test result. $h = 1$ means that H_0 is rejected at 5 % significance level (two-sided test): the ORRs of the two compared groups are statistically different. $h = 0$ means that H_0 could not be rejected at 5 % significance level.

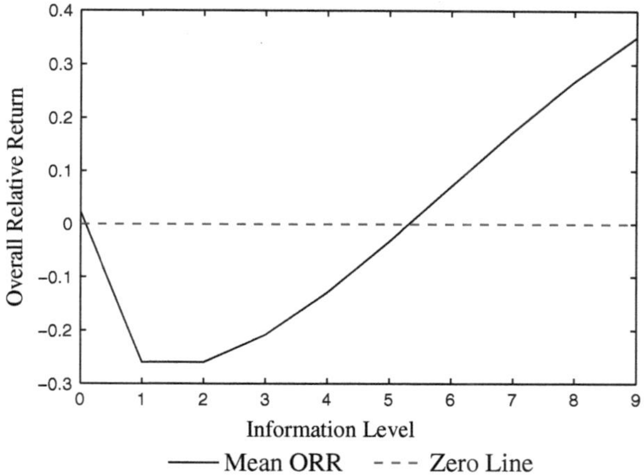

Fig. 1 Mean ORRs

to be true under the market efficiency hypothesis.[2] Moreover, his return is not statistically different from the return of the averagely informed agent I_6 whose information level serves as the border line between the "public information" and the "insider information" and as the break-even level between information levels providing negative and positive ORRs. The ORRs of the informed agents up to I_5 are negative. The information of all these agents is already included in the prices of the current period. The ORRs of the agents $I_1 - I_4$ are not proved to be significantly different to one another. All these agents lose in comparison with the I_0 or I_6. At the expense of these badly informed agents, the best informed agents $I_7 - I_9$ (or "insiders") beat the market. For these agents, the higher the information level is, the significantly higher the ORR will be.

The resulting J-curve has minor deviations from the curve presented in Huber (2007) which is based on the outcomes of an experimental market with human agents. The J-shape produced there suggests lower returns for the averagely- rather than for the low-informed agents. However, our results show that the averagely informed agents do not receive a return smaller than that of the least informed agent. Our results are in line with (Tóth and Scalas 2008).

We also run the simulation under a changed composition of the market population. If chartists are included, they become net losers. In addition, this change influences the position of the break-even point among the fundamental traders: the more chartists there are, the more informed fundamentalists manage to beat the market.

Furthermore, we discovered a market setup in which the J-curve is not valid. We change the share of random traders in the population (or the probability of

[2]We thank the anonymous referee for pointing this out. This is true if the random trader does not have any influence on the market price, or if he has an equal probability to beat the market and to be beaten by the market.

them being drawn in the market). With an extremely high random trader population, the J-structure is not observed, see Fig. 2. The random traders experience negative ORRs because they influence the price if their population is high, and they have a greater risk of being beaten by the informed fundamentalists. At the expense of the random traders, all of the informed traders utilize the information profitably; it corroborates with the predictions from Black (1986). On the other hand, when the number of random traders is relatively small, the benefits of the best informed traders are bolstered additionally by the losses of the least- and averagely-informed agents. Therefore, in Fig. 3 the mean ORRs of the best informed agent remain approximately unchanged, the ORRs of the random trader change slightly from positive to negative values, whereas the ORRs of the least informed trader increase drastically from negative to positive values when the share of random traders increases.

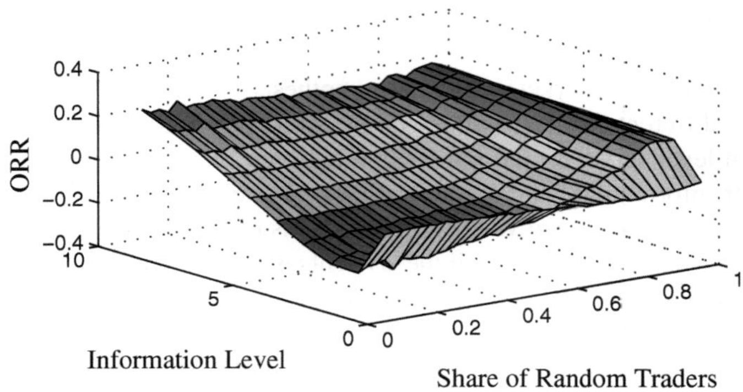

Fig. 2 ORRs and share of random traders: 3D representation

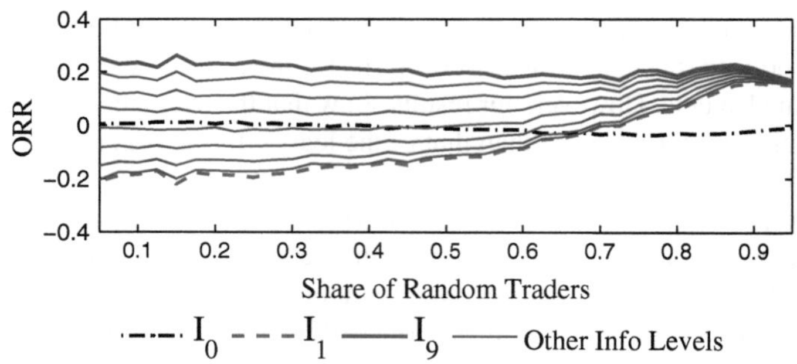

Fig. 3 ORRs and share of random traders: 2D representation per information level

4 Transaction Tax: Market Parameters and the J-Curve

Recently, the discussion about the necessity of transaction taxes has become topical once more. Transaction taxes are usually considered a typical regulation mechanism for stock markets, but the policy makers do not agree as to whether they have a positive or negative effect, and, consequently, upon the necessity of their implementation. The proposed Tobin tax varies between 0.005 % and 0.5 % depending on the country and security. Supporters believe that taxes may help to attain market stability. As a result, a share of the traders simply refrains from trading (Li et al. 2013). The argument against taxes is that it causes a decrease in trading volume. The impact on market efficiency and volatility is still being discussed (Hanke et al. 2010). We examine whether our model supports the results of the previous research and explore the ORR structure change along the tax level growth.

Our results show that transaction taxes negatively influence the trading volume as well as the acceptance ratio (parameters are defined as in Kirchler et al. 2011): see Fig. 4. However, small tax levels do not significantly affect these parameters. We examined normalized and mean relative absolute deviations as well as normalized returns and found no observable influence, which supports the results in Kirchler et al. (2011) and Umlauf (1993)[3] but contradicts the investigations in Li et al. (2013)[4] and Mannaro et al. (2008).[5] Pellizzari and Westerhoff (2009) propose that in a CDA, the stabilizing impact of transaction taxes is diminished and the volatility

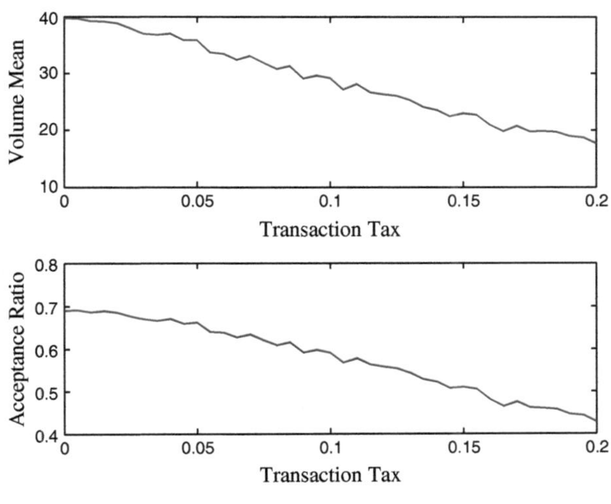

Fig. 4 Transaction taxes and market liquidity

[3]The results are based on the treatment of the real stock market in Sweden.

[4]According to their results, the volatility is reduced through the introduction of a higher tax level.

[5]They found that tax increases volatility and decreases trading volume.

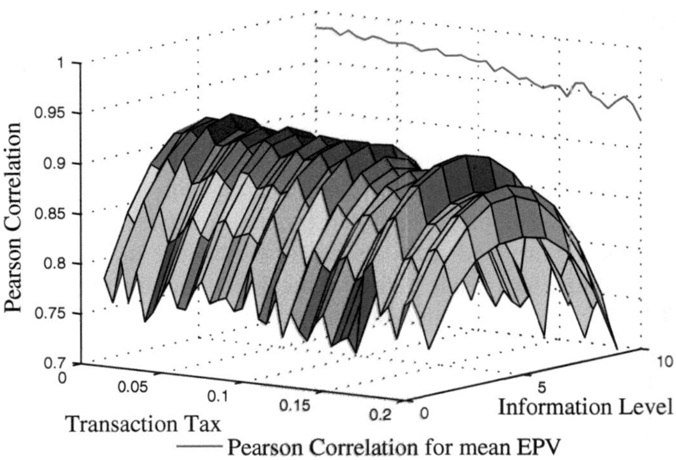

Fig. 5 Transaction taxes and the market efficiency

is not decreased, since a market liquidity reduction leads to a higher price impact of each order. On the contrary, the dealership market benefits from tax, since taxes deter speculative traders but the liquidity is still provided by specialists (dealers).

The liquidity decrease leads to another negative impact of transaction taxes, namely to a reduction in market efficiency: this supports the results from Posada and Hernández (2010). The market efficiency is measured by the Pearson correlation between the average prices per period and the EPVs. Figure 5 shows that the average price process has the highest correlation with the EPV of the agent with the median information level, which is in agreement with Huber (2007). It implies that the market prices do not reveal all available information. Only "public information", or the information of the least and averagely informed traders, is contained in the prices. The insiders' information is not reflected in the current prices, thus the market cannot be defined as one with "strong" efficiency. According to the classification of Fama (1970), our market shows "semistrong form" efficiency. Figure 5 also illustrates that high transaction tax rates lessen the degree to which prices reflect the amount of available information: the Pearson correlation becomes weaker. However, for very small transaction tax levels, the market efficiency is not critically affected, which confirms results in Kirchler et al. (2011).

Moreover, we find that an increasing tax level influences the structure of tax revenues received by the collecting institute from agents of different information levels: see Fig. 6. For a low level of transaction tax, the I_9-investor pays the minimal amount of tax, while the maximum amount is contributed by the I_1-investor. As tax increases, the relative tax revenues from the I_1 and I_9 increase, while those of averagely informed agents remain nearly unchanged.

Finally, with increasing tax rates, the inequality between information levels decreases: see Fig. 7. The J-curve is more pronounced if there is a small or zero transaction tax level.

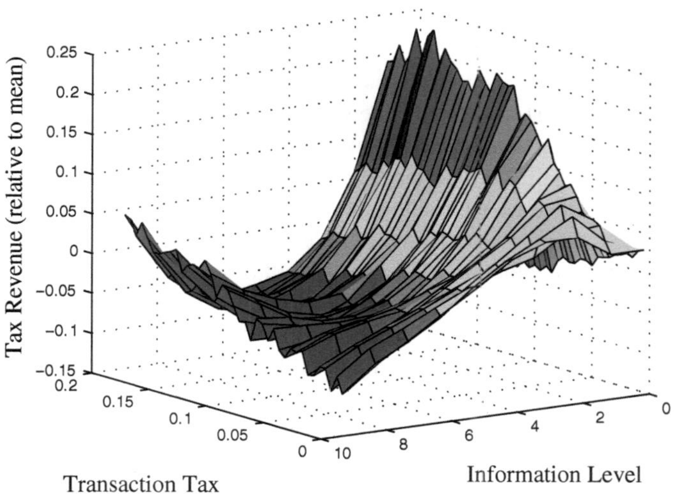

Fig. 6 Transaction taxes and tax revenues (the directions of tax axis and information level axis have been inversed for the purpose of better demonstration)

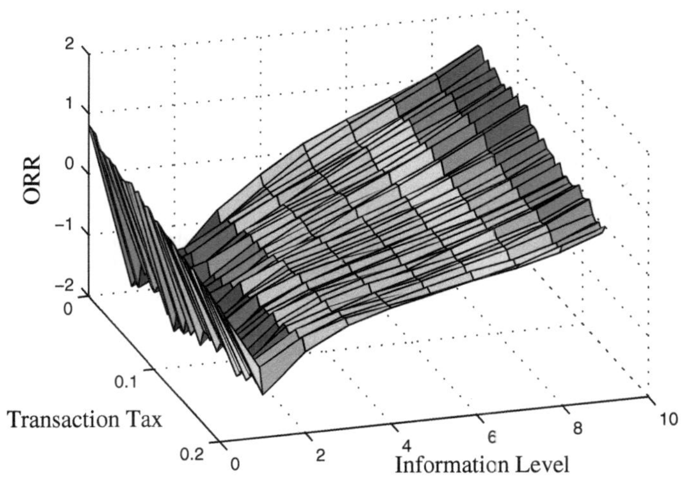

Fig. 7 Transaction taxes and ORRs (the directions of tax axis and information level axis have been inversed for the purpose of better demonstration)

Overall, in our simulated model, the market liquidity and efficiency decrease with tax growth, while the J-curve becomes less pronounced. However, at the level of transaction tax discussed in the literature, all of these changes are small.

Conclusion

This paper has provided new insights into the topic of relative wealth distribution amongst agents with heterogeneous information levels. For this purpose we reconstructed a computational ASM model, as presented in Tóth and Scalas (2008). The J-shape of the ORRs, mentioned in the previous studies, was confirmed with some modifications. We found that none of the low informed agents outperformed the least informed agent. We explored a counterexample in which the J-shape was not observed.

We extended the discussion by analyzing the J-curve reaction to the introduction of a transaction tax. The inequality between information levels became less critical with the introduction of taxes, while the market liquidity and efficiency were negatively influenced. However, the transaction tax levels considered in recent discussions had only marginal effects on the parameters discussed here.

References

Anufriev M, Arifovic J, Ledyard J, Panchenko V (2013) Efficiency of continuous double auctions under individual evolutionary learning with full or limited information. J Evol Econ 23:539–573

Bauwens L, Giot P (2001) Econometric modelling of stock market intraday activity. Advanced studies in theoretical and applied econometrics, vol 38. Kluwer Academic Publishers, Boston

Black F (1986) Noise. J Financ 41(3):529–543

Cervone R, Galavotti S, LiCalzi M (2009) Symmetric equilibria in double auctions with markdown buyers and markup sellers. In: Hernández Iglesias C, Posada M, López-Paredes A (eds) Artificial economics. Lecture notes in economics and mathematical systems, vol 631. Springer, Dordrecht, pp 81–92

Chiarella C, Iori G (2002) A simulation analysis of the microstructure of double auction markets. Quant Financ 2:346–353

Fama E (1970) Efficient capital markets: a review of theory and empirical work. J Financ 25(2):383–417

Grossman S, Stiglitz J (1980) On the impossibility of informationally efficient markets. Am Econ Rev 70(3):393–408

Hanke M, Huber J, Kirchler M, Sutter M (2010) The economic consequences of a Tobin tax - an experimental analysis. J Econ Behav Organ 74:58–71

Huber J (2007) 'J'-shaped returns to timing advantage in access to information - experimental evidence and a tentative explanation. J Econ Dyn Control 31:2536–2572

Kirchler M, Huber J, Kleinlercher D (2011) Market microstructure matters when imposing a Tobin tax - evidence from the lab. J Econ Behav Organ 80:586–602

Levy M, Levy H, Solomon S (2000) The microscopic simulation of financial markets: from investor behavior to market phenomena. Academic, San Diego

Li H, Tang M, Shang W, Wang S (2013) Securities transaction tax and stock market behavior in an agent-based financial market model. Proc Comput Sci 18:1764–1773

Mannaro K, Marchesi M, Setzu A (2008) Using an artificial financial market for assessing the impact of Tobin-like transaction taxes. J Econ Behav Organ 67(2):445–462

Pellizzari P, Westerhoff F (2009) Some effects of transaction taxes under different microstructures. J Econ Behav Organ 72(3):850–863

Posada M, Hernández C (2010) The effects of transaction costs on artificial continuous double auction markets. In: LiCalzi M, Milone L, Pellizzari P (eds) Progress in artificial economics. Lecture notes in economics and mathematical systems, vol 645. Springer, Berlin, pp 65–74

Posada M, Hernández C, López-Paredes A (2006) Learning in continuous double auction market. In: Beckmann M, et al. (eds) Artificial economics. Lecture notes in economics and mathematical systems, vol 564. Springer, Berlin, pp 41–51

Posada M, Hernández C, López-Paredes A (2006) Strategic behaviour in continuous double auction. In: Bruun C (ed) Advances in artificial economics. Lecture notes in economics and mathematical systems, vol 584. Springer, Berlin, pp 31–43

Schredelseker K (1980) Unequally distributed information and stock market theory. Arbeitspapiere des Fachbereichs Wirtschaftswissenschaft der Gesamthochschule Wuppertal 50

Schredelseker K (1984) Anlagestrategie und Informationsnutzen am Aktienmarkt. Schmalenbachs Zeitschrift für betriebswirtschaftliche Forschung 36:44–59

Tóth B, Scalas E (2008) The value of information in financial markets: an agent-based simulation. In: Hanke M, Huber J (eds) Information, interaction and (in)efficiency in financial markets. Linde, Wien

Tóth B, Scalas E, Huber J, Kirchler M (2007) The value of information in a multi-agent market model: the luck of the uninformed. Eur Phys J B 55:115–120

Umlauf S (1993) Transaction taxes and the behavior of the Swedish stock market. J Financ Econ 33:227–240

Westerhoff F (2010) A simple agent-based financial market model: direct interactions and comparisons of trading profits. In: Bischi G, Chiarella C, Gardini L (eds) Nonlinear dynamics in economics, finance and social sciences. Springer, Berlin, pp 313–332

What Is the Impact of Heterogeneous Knowledge About Fundamentals on Market Liquidity and Efficiency: An ABM Approach

Vivien Lespagnol and Juliette Rouchier

1 Introduction

The main goal of financial markets is to guarantee an optimal transfer of resources from supply to demand. This aim can be attained only if exchanges do actually occur in a considered period, i.e. the market is liquid. Amihud et al. (2005) write that "liquidity is a complex concept. Stated simply, liquidity is the ease of trading a security". Hence liquidity is a property of the system, which cannot be attained by just one agent with not enough influence on the market to create a context of easy trade. It is, however, an important feature which assures the functioning of the financial market through the behavior of the individual agents. It impacts the price, the volatility, and the amount of quoted orders.

During the last crisis, there was no way analysts could anticipate the liquidity and price falls that took place. Furthermore, to the best of our knowledge, there is no known mean to impact on the market liquidity. We are just faced with ex-post observations and attempt to understand the data. As an example: Air France-KLM's security was valued under 5€ on the CAC40, even if a consensus of analyst estimates that the book value was at least 6 times higher. However, nobody wanted to hold this asset so it was undervalued from the start! This is neither predictable nor rational. Actually, it has been shown that agent's behavior is drastically driven by asset exposure to liquidity risk (Amihud 2002). Liquidity is studied by micro structure theory, but it is usually taken as an exogenous parameter which influences the agents' behaviors. However, it has been little studied in

V. Lespagnol (✉) • J. Rouchier
Aix-Marseille School of Economics, Aix-Marseille University, CNRS & EHESS, Marseille, France
e-mail: vivien.lespagnol@univ-amu.fr

F. Amblard et al. (eds.), *Advances in Artificial Economics*, Lecture Notes in Economics and Mathematical Systems 676, DOI 10.1007/978-3-319-09578-3_9

agent-based computational economics, where it could be endogeneized as a result of actual transactions and taken into account by the agents.

The contribution of agent based economics is to produce models that integrate agents' bounded rationality, as well as their heterogeneity in terms of information and cognition. Several authors have already proven that this assumption of heterogeneity is necessary to reproduce, with models, results from actual behaviors (e.g. experimental data) (Bao et al. 2012).

The main goal of this paper is to focus on heterogeneous knowledge about fundamentals and its impact on liquidity dynamics in a financial market. We build an agent-based model, for which we make choices to produce the modeling structure and the rationality of agents. The comparison among different simulations shows that the information—that is available to different agents—has an impact on price dynamics and market liquidity. Different stylized facts are thus produced. The introduction of the belief perseverance (for the estimation of the fundamental value) enables to identify different bubble types: some that can be attributed to anchoring (Lord et al. 1979; Westerhoff 2004), and some that are generated by chartists behavior, based on trend extrapolation (Hommes 2006). As seen in other models (Giardina and Bouchaud 2003; Hommes et al. 2005; Lux and Marchesi 2000), we observe a destabilization power of chartists. We also witness the stabilization impact of the anchor on the price variance, since the trading price evolves more slowly than in the case of perfect knowledge of fundamentals. Finally, we test the aggressivity of orders (Parlour 1998) on the market efficiency (Fama 1970). As expected, if agents take into account the market liquidity as a parameter of price valuation, we will observe a rise in liquidity and a fall in efficiency. It would be possible to evaluate, with this mean, the price of liquidity in the system. However, at this stage we do not perform econometric tests, we just observe stylized facts.

2 Model

The model we base our market upon is Yamamoto's model (Yamamoto 2011). We extend it by adding a second trading asset. This two risky assets are assumed to be independent ($cov = 0$). Yamamoto defines the traders as heterogeneous in both their fundamentalist and their chartist ratios, their investment horizon and their risk aversion. The key parameters of this agent model are g_1 and g_2, which are—for each trader—generated at initialization following an exponential law of variance $\sigma_{g_1}^2$ and $\sigma_{g_2}^2$, respectively. A pure fundamentalist strategy has $g_2^i = 0$, whereas a pure chartist strategy has $g_1^i = 0$. When both values are higher than 0, the agent is a mix of both, which implies that he takes into account the chartist's and fundamentalist's expectations and make weighted average according to the g_1 and g_2 weight. From the two parameters values, the time horizon of investment and the risk aversion of each agent is also calculated. The more the agent tends to be fundamentalist, the more risk averse and long term investor she is. The converse is true: the higher the tendency to be chartist is, the lower risk aversion and the longer the investment horizon are.

Moreover, traders are boundedly rational. Indeed, their mood and their aggressivity influence the submission price. Concerning the types of orders, agents face: the market order (MO) and the limit order (LO). Finally, agents are not allowed to engage in short selling and are not monetarily constrained. For simplicity no quoted orders can be modify or cancelled. The trader who has submitted orders has to wait for its execution or for its limit execution date before submitting a new one.

In this paper, we distinguish two cases: a perfect knowledge one and a belief perseverance one. In both settings, we assume that the true fundamental value (f) follows a random walk and agents have access to the entire history of asset prices.

In case of perfect knowledge of the fundamental value, all fundamentalists have access to the good information. Because everybody has the same information and processes it correctly, the estimation of the fundamental value ($\hat{f}$) is assumed to be unique and right. For each fundamentalist i, it is mathematically expressed as:

$$\hat{f}_t^i = \hat{f}_t = f_t \quad , \forall i \tag{1}$$

In case of belief perseverance (imperfect knowledge of the fundamental value with adaptive learning), the forward fundamental value becomes idiosyncratic. It is a well known fact that traders make errors in their expectations and are overconfident (Barberis and Thaler 2003). Tversky and Kahneman (1974) highlight that people make estimation of prices by starting from an initial value that is adjusted across time. This is why we produce heterogeneity in our agents by giving them different initial believes: even if they get the same piece of information in time, they do not necessarily deduce the same fundamental value for the asset. Our formulation of the estimated fundamental value is inspired by Westerhoff (2004), so as to fit the anchoring assumption that is one aspect of bounded rationality. The fundamental value ($\hat{f}_t^i$) of agent (i) in case of belief perseverance is designed as:

$$\begin{aligned} \hat{f}_t^i &= \gamma_1\, p_{t-1} + \gamma_2\, \hat{f}_{t-1}^i + \gamma_3\, \hat{f}_{origin}^i \\ &\quad + N_t + a(\hat{f}_{t-1}^i - \hat{f}_{t-2}^i - N_t) \\ &\quad + b\,(f_{t-1} - \hat{f}_{t-1}^i) \\ \hat{f}_t^i &\neq \hat{f}_t^j \quad , \forall i \neq j \end{aligned} \tag{2}$$

The first line represents the anchor. It is defined by the last observed price (p_{t-1}), her previous expected fundamental value ($\hat{f}_{t-1}^i$) and her original expectation ($\hat{f}_{origine}^i$). γ_1, γ_2 and γ_3 represent the weight given to each component and add up to 1. The second line describes the anchor correction. The first component reflects the arrival of new information (N_t), common knowledge. The second component highlights the faith related to. As an example, if the recent update of the perceived fundamental value has been above the news impact ($\hat{f}_{t-1}^i - \hat{f}_{t-2}^i > N_t$), traders tend to overreact to news. The a parameter is the degree of misperception. This could be justify by the time needed to process it. The third line $(f_{t-1} - \hat{f}^i_{t-1})$—which is the spread between the last true fundamental value and the estimated one—is the one that represents learning. The b parameter affected to this learning is assumed to be close to zero. All parameters $(a, b, \gamma_1, \gamma_2, \gamma_3)$ are fixed and equal among traders.

The submission process is such that, at each time step, one randomly chosen agent goes through the four following steps.

1. for each asset available on the market, she formulates expectations on the forward price according to her features (g^i_1 and g^i_2).
2. she defines the amount of assets she wants to trade according to a CARA utility function.
3. she adapts her expectation according to her personal mood (simple rule of thumb). It is assumed to be time and asset dependent.
4. after having formulated an order, she corrects it according to the market depth and submits.

As a summary, the novelty of our model stays in the aggregation of an order book structure where two risky assets are traded and where fundamentalists do not have access to the true fundamental value. The mathematical model is developed in the Appendix.

3 Simulations Analysis

We have run 200 simulations for each trading round. A simulation is composed by 8,000 time steps. In order to exclude impacts of computer initialization, the first thousand time steps is excluded of the analysis. All in all, your market has run more than 256 trading rounds.

3.1 Dynamics in One-Type Markets: Fundamentalists

It has to be remembered that even agents who are all 100 % fundamentalists are not necessarily homogenous, since they differ in time horizon and aversion to risk. In our simulations, the original anchor is generated for each trader closely to the original true fundamental value ($\hat{f}^i_{origin} = f_0$). The anchor has a strong impact on the market. If it is not consistent with the fundamental value, and if all agents have approximately the same expectations, the market price will not reflect its fundamental. This is due to a slow learning process and a strong anchoring. The anchor also impacts the price oscillation. The variance falls from 696 to 554. Moreover, when the variance of the white noise of the true fundamental value changes ($\sigma^2_f = 0.1; 0.2; 1$), the variance of the market price is not really affected.

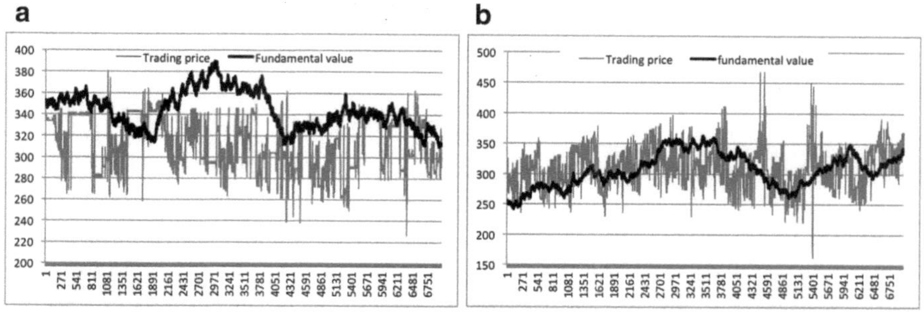

Fig. 1 Impact of learning speed in a one-type market with high variance of the fundamental value ($\sigma^2_{g_1} = 0.6, \sigma^2_{f_1} = 1, \sigma^2_{f_2} = 1$). (**a**) Asset 1: Low learning process (b = 0.005). (**b**) Asset 1: Fast learning process (b = 1)

In the original case (perfect knowledge of the fundamental), the fundamentalists perceive this movement and update their expectations according to the news. The price variance increases with the σ^2_f parameter. In the belief perseverance case, traders misunderstand the fundamental changes. Runs are more similar (decreasing standard deviation). Fundamentalists perceive the fundamental value more stable than it is, making the trading price more stable.

When the agents have their own believes about fundamental value (their own anchor), the trading price doesn't reflect the evolution of the true fundamental value (Fig. 1a). A way to correct this is to give more importance to the learning process, defined by Eq. (2), by increasing parameter b. This enables traders to update their expectations quickly, and makes the price more informative in case of high fundamental variance ($\sigma^2_{f_2} = 1$). The effect of the anchor is weaker, but still exists (Fig. 1b). When the true fundamental value falls, the trading price also decreases, but with a lag.

3.2 Dynamics in Two-Types Markets: Fundamentalists and Chartists

In addition to the belief perseverance, the traders are able to adapt their orders before submitting (Parlour 1998). In fact, the submitters have access to the five best quoted bids and asks (as some real markets). After formulating their expectations and before the submission, the traders can change the order type (market order, or limit order) and its limit price execution. Concretely, according to the market depth, the participants resolve the trade-off between accepting non-execution risk and paying the bid-ask spread. This choice that is made according to the visible part of the order book and the propensity of agent adaptation at time t has a direct impact on the market efficiency.

Table 1 Impact of components (with $\sigma_f^2 = 0.2$)

Parameter	$\sigma_{g_1}^2 = 0.6$	$+\hat{f}_i$	$\sigma_{g_1}^2 = 0.6, \sigma_{g_2}^2 = 1$	$+\hat{f}_i$	$+\beta = 2$
Average price	308.157	306.601	309.623	307.580	314.148
(Std. err)	(12.33)	(3.70)	(13.34)	(6.04)	(9.11)
Price variance	695.665	552.525	849.258	726.803	801.683
	(227.26)	(112.85)	(223.35)	(155.92)	(334.26)
Mean spread	7.79	6.93	9.21	8.16	14.70
	(4.58)	(11.74)	(6.09)	(12.30)	(15.80)
Spread interval	[−72; +77]	[−68; +77]	[−98; +132]	[−90; +115]	[−93; +141]
Overvalued	61.44 %	59.84 %	62.15 %	60.39 %	66.83 %
Liquidity	592.8899	573.1003	391.1416	392.1275	632.5406
	(92.22)	(33.32)	(26.55)	(21.38)	(76.86)
MO (buy)	20.17 %	20.64 %	20.55 %	20.77 %	16.62 %
MO (sell)	26.13 %	25.06 %	23.73 %	23.22 %	19.32 %

3.2.1 Perfect Knowledge of Fundamental Value

As we can expect—when a fundamentalist trend exists—the market never diverges even if chartists have a destabilizing impact. The average trading price in a one-type market—populated by fundamentalist ($\sigma_{g_1}^2 = 0.6$)—is 308.157 ECU. In the case of two-types market, the trading price is a little below 310 ECU, exactly 309.623 ECU in a market where agents are on average 43 % fundamentalist and 57 % chartists (a 43/57 % market). The destabilization power of chartists is present in the price variance (+22 %) and in the spread price (see Table 1).

In real world, agents are not myopic. They adapt their orders according to market observations. Traders change their behaviors and become more or less aggressive according to the market liquidity (fear of no-execution). We distinguish two cases, $\beta = 0$ and $\beta = 2$. With $\beta = 2$, agents are strongly influenced by the order book depth. The market liquidity is around 600, 1.5 times higher than the same market without order book looking (a 43/57 % market and $\beta = 0$). The mean price increases by 5 ECU compared to the same market without adaptive order ($\beta = 0$), and increases by 6 ECU compared to our previous one-type market. The maximum amount of assets traded in a time step is also hugely impacted. Agents adapt their submissions in order to increase the probability of order execution. The market price does not reflect the fundamentals, bubbles need times to burst (around 3,000 time steps). In a market with a null β, traders do not revise their expectations according to the market depth, the price oscillates around its fundamental value.

To sum up, adding chartists make the price oscillates more frequently and in a wide spread. The β parameter, because of arbitrage between higher price and no-execution risk, makes long term trend easily identified by chartists, and more persistent bubbles. The result is a non efficient market in any way. However, the β parameter increases the market liquidity.

3.2.2 Belief Perseverance

Relaxing the assumption of perfect knowledge of the fundamental gives us the same dynamics changes as in the one-type market. The mean trading price and its variance decreases compare to the perfect knowledge case (see Table 1). The news about fundamentals are misunderstood.

The anchor has a strong impact on the market dynamics. When the fundamental value moves, the anchor does not follow it. The result is a trading price far from the true fundamental. Let's take a closer look at Fig. 2b, d, the market price follows a double rise between time steps 800 and 2,500. The bubble occurs with a perceived fundamental value around 300 ECU and a true fundamental value around 200 ECU (Fig. 2d). The first increase corresponds to the difference between the estimated fundamental value and the trading price (attributed to chartist component). In the same time, a second bubble—based on the difference between the true fundamental value and the estimated one—appears (due to the belief perseverance of fundamentalists). The true fundamental value decreases while the estimated one stays relatively constant. Between the time steps 2,500 and 3,500, the price is relatively constant, but the true fundamental value increases (Fig. 2b). The bubble bursts partially. This price convergence is so independent of agents expectations.

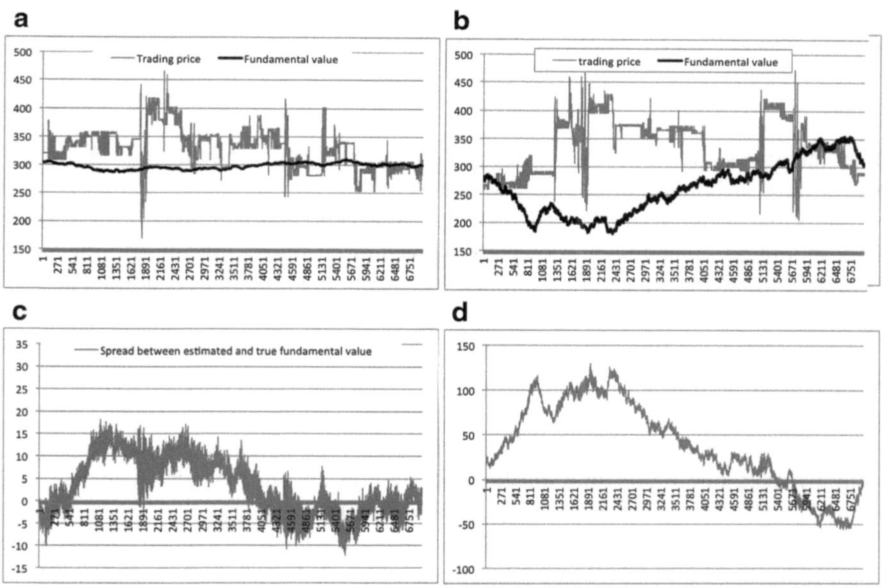

Fig. 2 Price dynamics with non-myopic agent in two-types market with belief perseverance ($\beta = 2$, $\sigma^2_{g_1} = 0.6$ and $\sigma^2_{g_2} = 1$). (**a**) Asset 1: Trading price ($\sigma^2_{f_1} = 0.2$). (**b**) Asset 2: Trading price ($\sigma^2_{f_1} = 1$). (**c**) Asset 1: Spread between the perceived fundamental value and the true one. (**d**) Asset 2: Spread between the perceived fundamental value and the true one

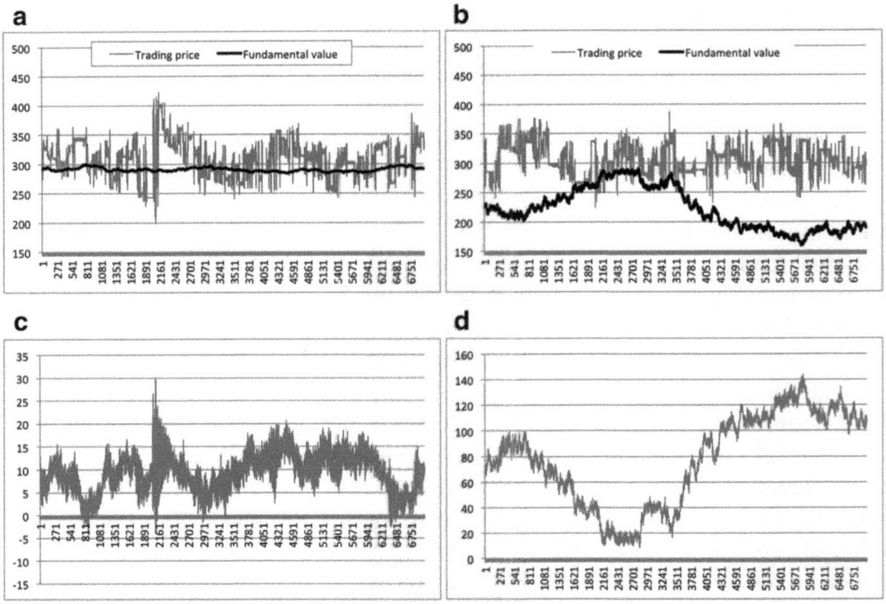

Fig. 3 Price dynamics with myopic agent in two-types market with belief perseverance ($\beta = 0$, $\sigma^2_{g_1} = 0.6$ and $\sigma^2_{g_2} = 1$). (**a**) Asset 1: Trading price ($\sigma^2_{f_1} = 0.2$). (**b**) Asset 2: Trading price ($\sigma^2_{f_1} = 1$). (**c**) Asset 1: Spread between the perceived fundamental value and the true one. (**d**) Asset 2: Spread between the perceived fundamental value and the true one

After the time steps 3,500, the trading price falls to the true fundamental value. This part of the convergence price is due to agents trading.

The same "double overvalued" price is observed even if traders do not adapt their submitted order function of the observed order book (Fig. 3b, d). The main difference is in the trend price implies by agents aggressivity. When agents are insensitive ($\beta = 0$), the trading price oscillates frequently around the fundamental (Fig. 3b). When agents are receptive to the order book statement, the market price can stay over- or under-valued for a longer period. The trends are easily observable, and the market less efficient. Concerning the market liquidity of this trading rounds, we are close to the liquidity volume of the one-type market. A rise in the chartist weight makes one more time the liquidity fall. A switch from 57 % to 81 % of chartists makes the volume decreases from 632 to 486.

As in the one-type market, the belief perseverance smoothes the market oscillations. It may explain inefficiency even in a fundamentalist market! Fundamentalists are able to generate endogenous bubbles.

Conclusion

In this paper we have built an order driven market, where two assets are traded, in order to study the liquidity and efficiency dynamics. We have proven that agents' types influence the market dynamics. Indeed, fundamentalists make the market oscillates around its fundamental price, while chartists make it diverge. Moreover, according to Beja and Goldman (1980), increasing the fundamentalist power in a previous stable system or adding chartists make the price oscillates in a higher spread and varies more but never diverges. In fact, chartists amplify fundamentalists' trend and make market inefficient. In case of a 43/57 % market, we have observed an increase of the average price by 2 ECU and of the variance by 22 % compared to the one-type market (fundamentalists). Moreover, adding chartists makes long life-duration bubbles, and so market less efficient. In any cases, this model is able to generate endogenous bubbles bloom and burst.

Concerning the liquidity, as we can expect, chartists increase the maximum amount of assets exchanged in a time step. Unexpectedly, they make the market less liquid than fundamentalists (on average). This is counter-intuitive to the findings of Shiller (2003) and Barber and Odean (2000). The propensity of agent to adapt her order to the market depth (β) permits to increase liquidity but decreases the market efficiency. Indeed, when traders adapt their order according to the observed order book, the market price diverges more easily. Because traders try to found a counter part, they care less about fundamentals. The trend are more observable with a hight β than a null one. This trends are driven by the fear of no-execution risk.

To assign bounded rationality to our agents, we relaxed the strong assumption of perfect knowledge of the fundamentals. Our adaptive agents estimate their own fundamental value according to a set of information (common knowledge), an anchor and a learning process. This produces decreasing price variance and smaller extreme values. If we stick to our definition of market efficiency as market price being close to the true fundamental value, market efficiency will be a function of the fundamental variance and the anchor. Indeed if the fundamental value has a high variance, the market will inefficient because of the belief perseverance. And if the anchor is close to the true fundamental value and this fundamental value has a low variance, the market will be more efficient than is the case of perfect knowledge of the fundamental. Adding to the usual bubbles that are created by speculators, we could produce bubbles due to belief perseverance of fundamentalists.

The next steps of this research are (1) to add noise traders so as to increase liquidity, (2) to add a social network for the spreading of information, (3) to improve the "estimated fundamental value" of our fundamentalists so as to distinguish quantitatively the speculative behavior from the misevaluating of the fundamental value.

Appendix: Mathematical Model

In this section, we focus on the mathematical equations of our order-driven market, where two risky assets are traded. It is based on a modified version of Chiarella et al. (2009) by Yamamoto (2011).

In our model, each trader is characterized by a fundamentalist component (g_1^i) and a chartist one (g_2^i). At period t, one randomly chosen trader formulates her expectation about the future return that will prevail in the interval ($t + \tau^i$).

$$\hat{r}^i_{t,t+\tau^i} = \frac{1}{g_1^i + g_2^i}\left[g_1^i . \ln\left(\frac{\hat{f}_t}{p_t}\right) + g_2^i . \bar{r}_t^i\right] \tag{3}$$

where τ^i is the investment horizon of agent i, and p_t denotes the spot price of the considering asset. The weights g_1^i and g_2^i are generated following an exponential law of variance $\sigma^2_{g_1}$ and $\sigma^2_{g_2}$, respectively. Note that a pure fundamentalist strategy has $g_2^i = 0$, whereas a pure chartist strategy has $g_1^i = 0$. The choice of a positive distribution is justified by the works of Hommes and Wagener (2009) and Hommes et al. (2007). They state that positive feedback for uninformed traders prevail in financial markets. By positive feedback traders, the literature means traders who buy and sell on momentum. Bao et al. (2012) state that with positive feedback traders, there is a self fulfilling oscillation around the fundamental value. Whereas with negative feedback traders, agents learn and make the price converge to its fundamental value.

The average stock return ($\bar{r}_t^i$) computed by chartists is defined by the expected trend based on the observations of the spot returns over the last τ^i time steps.

$$\bar{r}_t^i = \frac{1}{\tau^i}\sum_{k=1}^{\tau^i} r_{t-k} = \frac{1}{\tau^i}\sum_{k=1}^{\tau^i} \ln\frac{p_{t-k}}{p_{t-k-1}} \tag{4}$$

The forecasted return of the agent ($\hat{r}^i_{t,t+\tau^i}$) allows her to formulate the future expected price.

$$\hat{p}_{t+\tau^i} = p_t \exp(\hat{r}^i_{t,t+\tau^i}) \tag{5}$$

It has to be remembered that two risky assets are traded, therefore Eqs. (3)–(5) are applied to each one.

Once the expected prices are defined ($\hat{p}^1_{t+\tau^i}, \hat{p}^2_{t+\tau^i}$), the agent tries to maximize her utility function according to a budget constraint. We assume that the optimal demand of assets is defined by the maximization of a constant absolute risk aversion utility function (CARA) under a gaussian return of assets as:

$$\max_{W^i_{t+\tau^i}} \mathbb{E}^i_t[U(W^i_{t+\tau^i}, \alpha^i)] = \max_{W^i_{t+\tau^i}} \mathbb{E}^i_t[-exp(-\alpha^i.W^i_{t+\tau^i})]$$

$$W^i_t = z^{i,1}_t.p^1_t + z^{i,2}_t.p^2_t + C^i_t$$

where W^i_t reflects the agent's wealth and α^i her risk aversion. The wealth is composed by $z^{i,j}_t$ that denotes the amount of asset j owned by agent i at time t, p^j_t that is the spot price and C^i_t that is the cash invest in a risk free asset, like saving account or bond.

Regarding to the maximization, the optimal demand at the expected prices $(\hat{p}^{i,1}_{t+\tau^i}, \hat{p}^{i,2}_{t+\tau^i})$ may be expressed for asset 1 as:

$$\pi^{i,1}_t(\hat{p}^1_{t+\tau^i}, \hat{p}^2_{t+\tau^i}) = \frac{\ln\left(\frac{\hat{p}^1_{t+\tau^i}}{p^1_t}\right)}{\alpha^i \; p^1_t \; Var^i_1} \tag{6}$$

and for asset 2:

$$\pi^{i,2}_t(\hat{p}^1_{t+\tau^i}, \hat{p}^1_{t+\tau^i}) = \frac{\ln\left(\frac{\hat{p}^2_{t+\tau^i}}{p^2_t}\right)}{\alpha^i \; p^2_t \; Var^i_2} \tag{7}$$

The two assets are assumed to be independent ($Cov = 0$). In this case, $\pi^{i,1}_t$ is equal to the optimal demand of Yamamoto's paper (2011). The variance (Var^i_j) reflects the risk investment of asset j evaluated by agent i. It is assumed to be equal to the variance of the logarithmic of the return rate.

$$Var^i_j = \frac{1}{\tau^i} \sum_{k=1}^{\tau^i} [r^j_{t-k} - \bar{r}^{i,j}_t] \tag{8}$$

where $\bar{r}^{i,j}_t$ is the average spot return of asset j. A general writing is given by Eq. (4). The investment horizon (τ^i) and the risk aversion (α^i) are dependent of agent's features. We define them as Yamamoto (2011):

$$\tau^i = \tau \frac{1 + g^i_1}{1 + g^i_2} \tag{9}$$

$$\alpha^i = \alpha \frac{1 + g^i_1}{1 + g^i_2} \tag{10}$$

where τ and α are respectively a reference time horizon and a reference degree of aversion toward risk.

Finally, the agent has to define and submit her order. The amount of assets ($s_t^{j,i}$) the agent is willing to trade is determined by the absolute difference in the optimal demand for asset j at time t and $t-1$ as:

$$s_t^{i,j} = abs(\pi_t^{i,j} - \pi_{t-1}^{i,j}) \tag{11}$$

The sign of this difference ($\pi_t^{i,j} - \pi_{t-1}^{i,j}$), gives the agent's position. Notice that the submission price may differ of the expected one ($\hat{p}_{t+\tau^i}^{i,j}$, Eq. (5)) according to the agent's mood ($M_t^{i,j}$). The buy price is defined as:

$$b_t^{i,j} = \hat{p}_{t+\tau^i}^{i,j}(1 + M_t^{i,j}) \tag{12}$$

The sell price is:

$$a_t^{i,j} = \hat{p}_{t+\tau^i}^{i,j}(1 + M_t^{i,j}) \tag{13}$$

where $M_t^{i,j}$ is randomly assigned from a uniform distribution of mean zero.

References

Amihud Y (2002) Illiquidity and stock returns: illiquidity and stock returns: cross-section and time-series effects. Journal of Financial Markets 5:31–56

Amihud Y, Mendelson H, Perdersen LH (2005) Liquidity and asset prices. Found Trends Financ 1(4):269–364

Bao T, Hommes C, Sonnemans J, Tuinstra J (2012) Individual expectations, limited rationality and aggregate outcomes. J Econ Dyn Control 36:1101–1120

Barber BM, Odean T (2000) Trading is hazardous to your wealth: the common stock investment performance of individual investors. The Journal of Finance 55(2):773–806

Barberis N, Thaler R (2003) A survey of behavioral finance. Handb Econ Financ 1(B):1053–1128

Beja A, Goldman MB (1980) On the dynamic behavior of prices in disequilibrium. J Financ 35(2):235–248

Chiarella C, Iori G, Perello J (2009) The impact of heterogeneous trading rules on the limit order book and order flows. J Econ Dyn Control 33:525–537

Fama EF (1970) Efficient capital markets: a review of theory and empirical work. J Financ 25 (2):383–417

Giardina I, Bouchaud J-P (2003) Bubbles, crashes and intermittency in agent based market models. Eur Phys J B 31:421–437

Hommes C (2006) Heterogeneous agent models in economics and finance. Handb Comput Econ 2:1109–1186

Hommes C, Wagener F (2009) Complex evolutionary systems in behavioral finance. In: Hens T, Schenk-Hoppe K (eds) Handbook of financial markets: dynamics and evolution. Elsevier, pp 217–276

Hommes C, Huang H, Wang D (2005) A robust rational route to randomness in a simple asset pricing model. J Econ Dyn Control 29:1043–1072

Hommes C, Boswijk P, Manzan S (2007) Behavioral heterogneity in stock prices. J Econ Dyn Control 31:1938–1970

Lord C, Ross L, Lepper M (1979) Biased assimilation and attitude polarization: the effects of prior theories on subsequently considered evidence. Journal of Personality and Social Psychology 37:2098–2109

Lux T, Marchesi M (2000) Volatility clustering in financial markets: a micro-simulation of interactive agents. International Journal of Theoretical and Applied Finance 3:675–702

Parlour CA (1998) Price dynamics in limit order markets. Rev Financ Stud 11(4):789–816

Shiller RJ (2003) From efficient markets theory to behavioral finance. The Journal of Economic Perspectives 17(1):83–104

Tversky A, Kahneman D (1974) Judgment under uncertainty: heuritics and biases. Science 185(4157):1124–1131

Westerhoff FH (2004) Multi-asset market dynamics. Macroecon Dyn 8:596–616

Yamamoto R (2011) Order aggressiveness, pre-trade transparency, and long memory in an order-driven market. J Econ Dyn Control 35:1938–1963

An Agent Based Propagation Model of Bank Failures

André Dias, Pedro Campos, and Paulo Garrido

1 Introduction

Financial crisis and bank failures have been topics of interest for many researchers (Friedman and Schwartz 1963; Bernanke 1983; Bernanke and Gertler 1989). The recent crisis of 2008 prompted for scientific developments in the field increasing the contributions and new approaches on the subjects. The present work proposes a new paradigm to the modeling of a banking network in what concerns the assessment of its resilience in the presence of bank failures. For this purpose a model is proposed where banks interconnect through the inter-banking market and consumers' decisions. Our main goal is to assess how the banking structure handles credit and liquidity shocks. Netlogo (Wilensky 2013), a multi-agent programmable modeling environment, is used to create the agent-based banking network model, to simulate the effect of shocks, and to collect the results.

A simulation is started by an external shock causing one bank to fail with the goal to trigger the cascade effect of network contagion. The analysis is a very short-run and a rapid one, as we limit our model to the verification of the banking network resilience. We have used three different types of agents—banks, consumers and a central bank, each with its own individual characteristics—in the network model. Simulations are initiated by shocks from the asset side and from the liabilities side of balance sheets created in a *scale-free* network (Georg 2011).

A. Dias (✉) • P. Campos
LIAAD INESC TEC and Faculdade de Economia, Universidade do Porto, Porto, Portugal
e-mail: DiasAndre@gmail.com

P. Garrido
Escola de Engenharia, Universidade do Minho, Braga, Portugal

F. Amblard et al. (eds.), *Advances in Artificial Economics*, Lecture Notes
in Economics and Mathematical Systems 676, DOI 10.1007/978-3-319-09578-3_10

2 Literature Review

Literature on financial crises is abundant. Several recognized economists, like Milton Friedman (Friedman and Schwartz 1963) or Ben Bernanke (Bernanke 1983; Bernanke and Gertler 1989) have produced research relating bank failures with financial crises. The Great Depression (GD) was their main inspiration, as it was understood that if policy was successful in avoiding a GD, then the financial system should be safe. Between 1929 and 1933 the United States of America saw the number of banks decline from over 25,000 to less than 15,000 (Friedman and Schwartz 1963). The 2008 crisis showed a relationship between bank failures and financial crises similar to the GD. It is commonly referred that the 2008 crisis was triggered by the failure of the Lehman Brothers and Bear Stearns investment banks. The reduction in confidence originated a severe shrinking of bank lending, affecting all industries including banking itself (Ivashina and Scharftein 2009; Blavarg and Nimander 2002), leading to the decrease in economic activity and thus driving the world to a financial crisis.

Bank's balance sheets analysis is the starting point in banking networks simulation models (Chan-Lau 2009; Chan-Lau et al. 2009). The main inspiration for many authors is the model of Allen and Gale (2000), as they presented one of the first models of a banking network, and got to the conclusion that system resilience depends on the structure of relationships. Further studies on network contagion have been produced. Some focus on the asset side of the balance sheet,—credit shocks— like (Nier et al. 2007) who studied the resilience of the network in case of a general loss of assets among all the banks in the system. Also, Georg (2011) studies the ability of a central bank to maintain stability.

Other studies focus on the liability side of the balance sheet—liquidity shocks— like (Diamond and Dybvig 1983) who studied the main function of a bank, which is transforming maturities of deposits. Liquidity shocks happen when banks funding sources are at risk. Market stress levels or *fire-sales* have also been one topic of interest (Allen and Gale 1998). Another topic worth mentioning is that financial markets have grown more complex with the sort of financial instruments that are now built in the form of derivatives. It has been pointed by authors as (Markose et al. 2012) that their complexity was one of the reasons why so many players were unaware of the possible devastating consequences in case of default.

Literature on financial crises and banking failure is quite vast, and our model obtains inspiration from very different perspectives, such as Agent-Based modeling, which remains an unusual approach to deal with this kind of problems.

3 The Model

The model is based on a network, representing banks, consumers and their relations. A node can represent a bank or a consumer. Links connecting two bank nodes represent a credit and debt relationship existing between them (see Fig. 1) Links

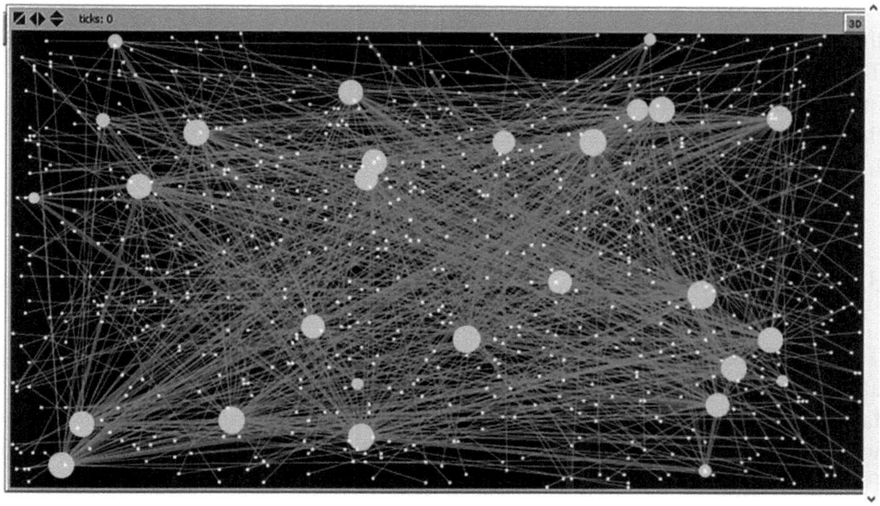

Fig. 1 Banking Network Representation. *Light grey nodes* represent banks; *white small nodes* represent consumers. *Darker lines* connect banks to banks; *grey lines* connect consumers to banks

connecting a consumer and a bank represents a deposit of the consumer in the bank to the amount of one monetary unit. We consider the banking network to include the whole of the bank-to-bank links and the consumers to bank links. Netlogo is used to create the nodes and links. Links are created in a way that ensures the network is scale-free.

Banking shocks are deeply connected with the concept of bank failure. Usually bank failures occur for two main reasons: when banks are unable to recover lent money (known as credit shocks); and when a loss of deposits, their main source of funding, occurs (known as liquidity shocks). Both these situations lead to losses that ultimately have to be supported by bank's capital and in case this is not sufficient to support the loss, the bank defaults.

Credit shocks are many times called asset side shocks and liquidity shocks are many times called liability side shocks. The reason for this to happen is that credit shocks appear when a bank must reduce its asset side of the balance sheet due to the default of the bank to which it has lent money. In this type of loss, the bank's capital is diminished to the amount of the loss.

On the other hand, liquidity shocks are related to withdrawals of funds. As the asset side of the balance sheet is supported by the financing sources registered on the liability side, if the funds are not replaced there will be a balance sheet contraction. In a normal market situation the amount of assets the banks will have to sell is equal to the amount of funds lost. This way, capital will not be diminished and, actually, the solvability ratio will even increase resulting in a lower probability of banking default (Chan-Lau 2009).

If normal markets conditions are not present, we will have a fire sale meaning that the market will price assets below their value causing a loss to the bank's capital. This usually happens when there are liquidity shortages, the assets are too illiquid or a big amount of the same type of asset is being sold (Hagan 2009). Either way, the bank will be forced to sell a bigger amount of assets than the amount of funds lost.

3.1 Agents and Rules

There are three main rules of agents' behavior. They are described in Table 1. Banks, after each contagion cycle must check their capital level. If this is negative they will fail and exit the network. Consumers with deposits in surviving banks decide if they wish to continue in the network or leave by withdrawing the deposit. The Central Bank is the lender of last resort on the condition that a bank satisfies the minimum required capital value to receive a loan.

In the table, a fourth agent, formally named "Meta", embodies the rule to stop a simulation: if no bank has seen its capital to become negative then propagation of bank failures ended and there is no point in keeping the simulation running.

The formalization of the model starts with the bank's balance sheet identity: $Assets = Liabilities + Capital$. The balance sheet of bank b_i will have the inter-

Table 1 Main rules of agents' behavior

	Agent	Rule	Behavior
1	Bank i	Run a capital check	If capital $i < 0$ after Central Bank inspection then "Bank i leaves the network"
2	Consumer j	Remaining in the network	If random $j >$ probability of exiting the network, then "Consumer j leaves the network"
3	Central Bank	Inspect if loan condition applies	If capital $i < 0$ then {if capital $i >$ admissible value, then "Loan is granted"}
	"Meta"	Stop simulation	If for all i capital $i >= 0$ then "Stops simulation"

banking loans given to other banks $IB^C{}_i$, illiquid investments Iv_i[1] and legal obligations LO_i, on its assets side. On the liability side of bank b_i balance sheet there are the clients' deposits D_i, the borrowing positions $IB_i{}^D$ and capital C_i. The bank b_i balance sheet is given by (1).

$$IB_i{}^c + Iv_i + LO_i = D_i + IB_i{}^D + C_i \tag{1}$$

The total amount of loans conceded by bank b_i in the inter-banking market at moment t is giving by (2) where π_{ij} represents the amount bank b_i lent to bank b_j.

$$IB_i^c(t) = \sum\nolimits_j \pi_{ij}(t) \tag{2}$$

We assume that, at the start of a simulation, all banks borrowing from bank b_i get the same amount of money. A fraction of the client's deposits $D_i(t)$ of amount μ goes to the inter-banking market. The number of banks borrowing from bank b_i is given by n.

$$\pi_{ij}(t) = \mu D_i(t)/n \tag{3}$$

From the observation of the network and with the knowledge of the π_{ij} for each bank, we are able to obtain the bank's borrowing position $IB_i{}^D$:

$$IB_i^D(t) = \sum\nolimits_j \pi_{ji}(t) \tag{4}$$

Capital C and legal obligations LO must follow the following set of rules, namely a solvability ratio ω and a legal reserves ratio Ψ[2]:

$$\begin{gathered} \omega \le \frac{C_i}{IB_i^C(t) + Iv_i(t) + LO_i(t)} = \frac{C_i}{IB_i^D(t) + D_i(t) + C_i(t)} \\ \Psi \le \frac{LO_i(t)}{D_i(t) + IB_i^D(t)} \end{gathered} \tag{5}$$

A most important role of a central bank is the stabilization of the financial system under its control. One way to accomplish this goal is by being the lender of last resource (Georg 2011). The inclusion of the central bank in the model aims at

[1]For illiquid investments, we understand investments on firms, mortgages or even expensive goods. The source of funds that supports these investments is not dependent on the consumers presented in this model.

[2]These ratios are assumed to be followed at the start of the simulation indicating an equilibrium situation. However, during the ongoing of the simulation they might not be followed because we are making a very short-run analysis.

dealing with situations where banks are in a default position. In this situation, we assume that the central bank will inject money in the default bank, raising the value of IB_i^C.

Equation (6) shows that the central bank is willing to loan to the defaulting bank either the negative capital amount *NF* or a fraction *a* of the illiquid investments, whichever is less. The money lent by the central bank will enter the bank's capital in the asset side as IB^C. With this loan the bank will remain in the network and will continue to operate.

$$BC_i^D(t) = \min\left(NF(t), a^* Iv_i(t)\right) \tag{6}$$

The idea of *moral hazard* (Dembe Allard and Boden Leslie 2000) was introduced in the model by increasing the risks of banks when they enter the inter-banking market. To keep the model simple—remembering that we do not take into account the return on the illiquid investments—moral hazard is modeled by creating the possibility that a bank can transfer amounts of money from *LO* to IB_i^C failing to hold Ψ at the start of the simulation. By putting this amount available in the inter-banking market, the risk of bank failure in case of a credit shock is increased, because increased borrowed amounts might not be recovered. If this happens, there will be an increased need to sell illiquid investments relatively to the situation where the banks follow Ψ from Eq. (5), resulting on a greater need of capital to keep the balance sheet solvent.

All clients' deposits *D* are on demand. The probability $P(W)$ of a client withdrawing his money from a bank depends on the ratio between deposits in time *t* and initial consumers *CS* times a factor *L* that represents the sum of an individual choice function ρ and a group choice function Γ. The expression for $P(W)$ is given in Eq. (7). Withdrawals are simulated at the end of the each simulation step (Kirman 1993).

$$P\left[W(t)\right] = \frac{D(t)}{CS}L = \frac{D(t)}{CS}\left[\frac{\rho}{CS} + \Gamma\left(1 - \frac{D(t)}{CS}\right)\right] \text{ for } t > 0 \tag{7}$$

A bank's net position $S_i(t)$ is measured by its capital. This is a vital simulation check to evaluate a bank's continuation in the network. If bank's b_i net position is $S_i(t) \geq 0$, then the bank is solvent and it will continue in the network. On the other hand, if $S_i(t) < 0$ a bank is in a default situation and it will be removed from the network unless it receives a loan from the central bank. The loan from the central bank will be equal to:

$$NF_i(t) = S_i(t), \qquad \textit{if } S_i(t) < a \cdot Iv_i(t) \tag{8}$$

Between moment *t* and moment $t+1$ there is an intermediate step. This is the moment when banks balance sheets are updated due to the credit shock at time *t*. All the banks affected will see their balance sheets categories updated with the exception of Iv_i and IB_i^D.

$$\overline{IB_i^C} + Iv_i + \overline{LO}_i = \overline{D}_i + IB_i^D + \overline{C}_i \tag{9}$$

After this intermediate step, we will update the balance sheet due to the liquidity shock. This will provide us with the initial balance sheet for moment $t + 1$.

$$IB_{i(t+1)}^C + Iv_{i(t+1)} + LO_{i(t+1)} = D_{i(t+1)} + IB_{i(t+1)}^D + C_{i(t+1)} \tag{10}$$

The selling of illiquid investment Iv_i in a crisis situation is very likely to happen without normal market conditions. As such, buyers will buy these assets at a discount price. This will produce a loss that must be supported by the bank's capital. So if bank b_i loses X of inter-banking funds, this will cause a loss in Iv_i of amount $(1 + \delta)X$ causing a reduction on the banks b_i capital of δX. We can see δ as the stress level in the illiquid assets market.

3.2 Simulation Parameters and Steps

A simulation run is started by a (exogenous) credit shock, provoked by the user. This means that a chosen bank is made insolvent by adequately changing variables in its balance sheet, according to Eq. (1). Thereafter, a simulation cycle is made where all the banks are visited to verify if their net position $S_i(t)$ stays non-negative; if that is not the case, the bank is removed. Simulation cycles are repeated until no bank appears with a negative net position. If so, the simulation run is terminated, as the effects of the shock or the cascade propagation of failures has been determined for the simulated situation.

The model was written in Netlogo, and we setup nine different parameters whose values were divided in four groups, in order to better understand the effects of different entities—market, banking authorities, banks, consumers—in the banking network. A population of 30 banks and 700 consumers was created and the model was run in each configuration for 100 times. Every model initialization creates a slightly different network for the same parameters values due to the stochastic mechanism created to link nodes. Simulations were run for each group, altering their controlled parameters for the allowed range and for every possible combination, while the remaining parameters stayed constant. To increase the scope of results analysis we ran the simulation as described above, first by selecting the biggest bank of the network in terms of capital to default first and then a second time selecting the smallest bank in terms of capital to default first (Table 2).

Table 2 Model parameters divided by groups (lines "Group 1" to "Group 4")

				Simulation setting			
Group	Parameters	Function	Range	Grp 1	Grp 2	Grp 3	Grp 4
1	Skew degree	Level of skewness of the network	[0; 1]	[0; 1]	0.5	0.5	0.5
	Fire sale	Level of stress of the Iv_i market		[0; 1]	0.4	0.4	0.4
2	LO ratio	Sets the reserves requirements		0.1	[0; 1]	0.1	0.1
	Solvability ratio	Sets the solvability requirements		0.3	[0; 1]	0.3	0.3
	Central Bank coefficient	Percentage from Iv_i that central banks accept as collateral		0.2	[0; 1]	0.2	0.2
3	Inter-Bank market coefficient	Percentage taken from deposits to the Inter-Bank market		0.6	0.6	[0; 1]	0.6
	Moral Hazard coefficient	Percentage from LO to IB_i^C		0.5	0.5	[0; 1]	0.5
4	Group Choice coefficient	Withdrawal recruitment		0.1	0.1	0.1	[0; 1]
	Individual Choice coefficient	Individual recruitment choice		0.03	0.03	0.03	[0; 1]

The first group corresponds to market controlled parameters; the second corresponds to banking authorities controlled parameters; the third corresponds to parameters controlled by banks; the fourth group includes consumers controlled parameters. When a simulation runs for a specific group, parameters from that group change within the presented range while all the others remain constant

4 Results

We have focused the collection of results in terms of banks survival ratio *SR* given by the number of surviving banks, *Sb*, over the total banks in the beginning of the simulation, *Ib*:

$$SR = \frac{Sb}{Ib} \tag{11}$$

Systemic risk was found for parameter groups 1 and 2. For group 1, when selecting the biggest bank in terms of capital to default first, we concluded that the higher the values of the Fire-Sale and Skew Degree coefficients, the smaller is the number of surviving banks. For values of both parameters above 0.9, no banks survive and so the survival ratio is 0. Changing the Fire-Sale coefficient alone does not create a situation of systemic risk; for values below 0.3 the network is resilient. However, from simulations it became clear that the higher are the levels of stress in the markets for illiquid assets, the smaller are the chances of banks surviving shocks. On the other hand, when selecting the smallest capital bank present in the network to default first, the higher the Skew Degree the higher the survival ratio is (Fig. 2).

Analyzing the results for group 2 with the biggest bank defaulting first, we obtained systemic risk if the LO Ratio and the Solvability Ratio go very low. In fact, if both parameters go below 0.1, the survival ratio is 0. We can see a very sudden drop of surviving banks, when both parameters go below the threshold of 0.4. Results stay similar for different values of Central Bank coefficient and also for choosing the smallest capital bank in the network to start the crisis.

For group 3 we were not able to find systemic risk even for extreme parameters settings. Nevertheless, we clearly found a downward trend in the survival bank ratio when both the Inter-Bank Market coefficient and the Moral Hazard coefficient increased in the situation where the biggest capital bank defaults first. For parameter values approximately 0.5 and higher the survival ratio became flat. When the smallest capital bank is chosen to default first, there was no contagion to report (Figs. 3 and 4).

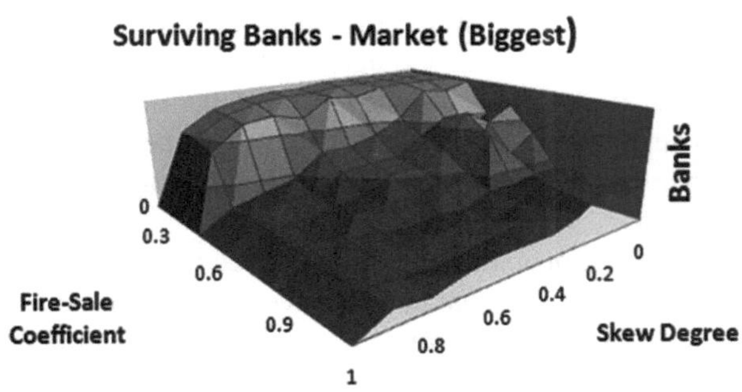

Fig. 2 Surviving banks for group 1 when the biggest bank is selected first to default

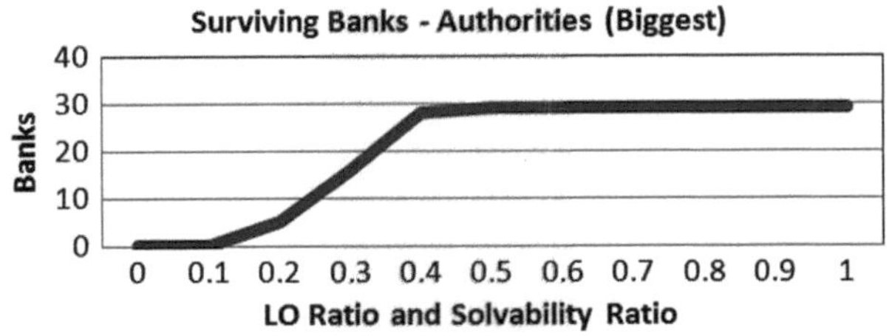

Fig. 3 Surviving banks for group 2 when LO Ratio and Solvability Ratio change one by one and when selecting the biggest bank to default first

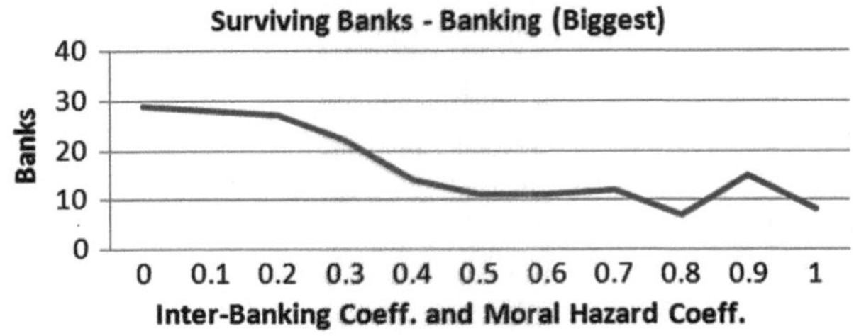

Fig. 4 Surviving banks for group 3 when the Inter-Banking coefficient and the moral hazard coefficient change one by one, selecting the biggest bank to default first

Table 3 Minimum and maximum values observed for the survival ratio per group

	Survival ratio intervals (biggest bank to default first)
Group 1	[0; 1]
Group 2	[0; 1]
Group 3	[0.23; 0.96]
Group 4	[0.23; 0.73]

For group 4 there was also no presence of systemic risk and the default ratio was very stable not showing any presence of a trend that would make us believe that credit shocks are more devastating than liquidity shocks. Essentially, both types of shocks are present in each run of the simulation. Nevertheless, when we amplified the liquidity part of the shocks we were not able to observe systemic risk (Table 3).

Another result we could observe and agree with Georg (2011) and Provenzano (2013), is that a scale-free network, which has a smaller number of linkages when compared with other theoretical networks, often produces isolated banks in the network. Isolation prevents the banks to infect or being infected. They are usually the survivors of a crisis.

As an example of the explanation above we can see on the left of Fig. 5, the dark dots representing the three banks that have survived the default of the first bank and that have no remaining linkages with other banks. At the end of the simulation, part of the network was resilient to the initial shock and the surviving banks are exactly

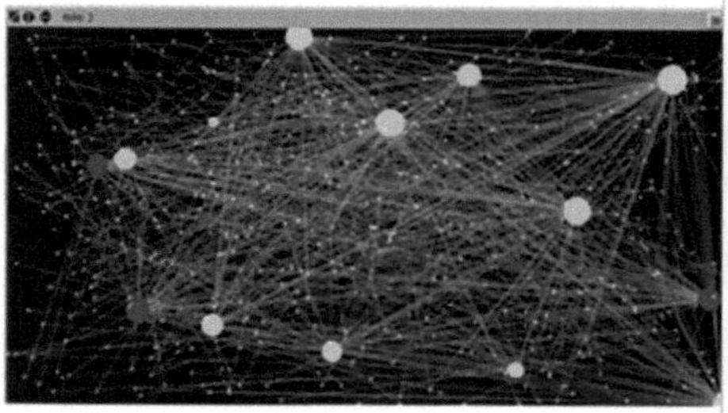

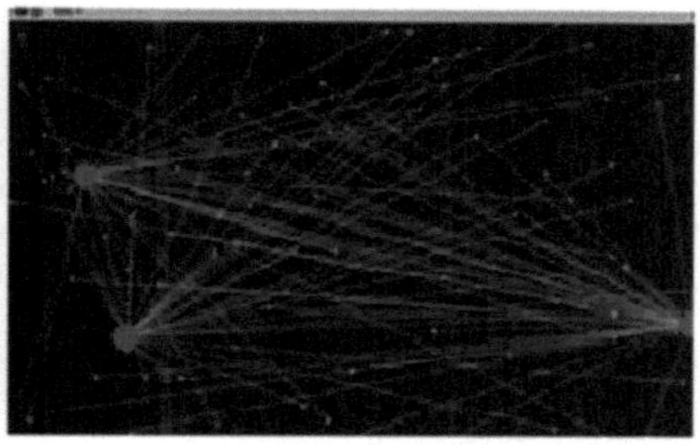

Fig. 5 Banking Network after first impact (*left*); network after simulation ends (*right*)

the ones that got isolated since the first run of impact, as we can see on the right side of Fig. 5.

5 Final Remarks

Our model combines different characteristics referred in the literature and aims at creating a banking network model that better reflects reality. For this purpose, three main points were taken into consideration: first, we used *three* different types of agents: banks, consumers and a central bank; second, we modeled the possibility of *both* credit and liquidity shocks; and third, we used a *scale-free* network topology, the most commonly observed in banking networks.

Using an agent-based approach, we defined micro behaviors for the three types of agents, whose possibilities of evolution along the unfolding of the simulation were controlled by several parameters. Changing the values of parameters, we were able to study the resilience of the network in different conditions.

Results showed the presence of systemic risk for two groups of parameters. They reinforce the existing evidence of real networks being scale-free. Taking this into account, we vow to contribute to the understanding of systemic risks in banking networks and help policy makers on their road to produce measures that keep financial systems stable.

In terms of policy, simulations supported the idea that banking systems should not become too concentrated, because a crisis triggered by the biggest bank will produce devastating results on the whole network. On the other hand, when the smallest bank triggers the crisis, increasing the Skew Degree will decrease the chances of systemic risk. Also, results of simulations support the view that authorities should keep straight rules in terms of reserve ratios and solvability ratios minima. Too low ratios in situations of high stress levels in the markets can cause banks to be left unprotected to abrupt events in the banking network and without options to survive.

Acknowledgments The work of author Pedro Campos is financially supported by FCT—Fundação Portuguesa para a Ciência e a Tecnologia/MEC—Ministério da Educação e Ciência through national funds (PIDDAC) and the ERDF—European Regional Development Fund through ON2 North Portugal Regional Operational Programme within project "NORTE-07-0124-FEDER-000059." The work of author Paulo Garrido was financially supported by FCT, Fundação Portuguesa para a Ciência e a Tecnologia, through the Program PEst, Strategic Program of the Algoritmi Research Center, Project FCOMP-01-0124-FEDER-022674.

References

Allen F, Gale D (1998) Optimal financial crises. J Financ 53(4):1245–1248

Allen F, Gale D (2000) Financial contagion. J Polit Econ 108(1):1–34

Bernanke BS (1983) Nonmonetary effects of the financial crisis in propagation of the great depression. Am Econ Rev 73:257–76

Bernanke BS, Gertler M (1989) Agency costs, net worth, and business fluctuations. Am Econ Rev 79:14–31

Blavarg M, Nimander P (2002) Interbank exposures and systemic risk. In: Risk measurement and systemic risk, Bank for International Settlements, Basel

Chan-Lau JA (2009) The global financial crisis and its impact on the Chilean banking system, mimeo. International Monetary Fund, Washington

Chan-Lau JA, Espinosa M, Giesecke K, Solé J (2009) Assessing the systemic implications of financial linkages, Chapter 2. In: Global financial stability report, International Monetary Fund, April 2009

Dembe Allard E, Boden Leslie I (2000) Moral hazard: a question of morality? New Solut 10(3):257–279

Diamond D, Dybvig P (1983) Bank runs, deposit insurance, and liquidity. J Polit Econ 91(3):401–419

Friedman M, Schwartz AJ (1963) A monetary history of the United States. Princeton University Press, Princeton, pp 1867–1960

Georg C-P (2011) An agent based approach to interbank networks and monetary policy, working paper, Jena University

Hagan WD (2009) House of cards: a tale of hubris and wretched excess on Wall Street. Doubleday, New York

Ivashina V, Scharftein D (2009) Bank lending during the financial crisis of 2008. J Financ Econ 97(3):319–338

Kirman A (1993) Ant, rationality, and recruitment. Q J Econ 108(1):137–156

Markose S, Shaghaghi AR, Giansante S (2012) Too interconnected to fail financial network of US CDS market: topological fragility and systemic risk. J Econ Behav Organ 89(3):627–646

Nier E, Yang Y, Yorulmazer T, Alerton A (2007) Network models and financial stability. Working paper, Bank of England

Provenzano D (2013) Contagion and bank runs in a multi-agent financial system. In: Managing market complexity, lecture notes in economics and mathematical systems, vol 662. Springer, Berlin, pp 27–38

Wilensky U (2013) NetLogo, Center for Connected Learning and Computer-Based Modeling, Northwestern University, Evanston. http://ccl.northwestern.edu/netlogo/

Direct vs. Side Effects in Financial Contagion: What Weights More?

Stefano Zedda

1 Introduction

As defined by Basel Committee on Banking Supervision (2010), robust financial systems are those that do not adversely induce the propagation and amplification of disturbances affecting their soundness. As a consequence, a key issue for financial supervision and macro-prudential regulation is understanding and quantifying the links between financial and public sectors.

An important reference for describing the contagion channels is the IMF (2010) representation (set out in Fig. 1) of the interconnections between banks and sovereigns in a domestic and international perspective, even if it has not yet be translated into a formal model and tested.

In literature, many of these effects have been considered one at a time.

The direct influence of bank riskiness on public finances is mainly represented by the contingent liability for State support to the banking sector in case of distress. State aid often shows up trough an injection of funds to a distressed banking sector: In the period between 1 October 2008 and 1 October 2011, the European Commission approved aid to the financial sector for an overall amount of EUR 4.5 trillion (36.7 % of EU GDP).

Some quantifications of this risk are in European Economy (2011), where an ex ante perspective is used for evaluating the costs of banking crises on national accounts.

The positive relationship between public debt and interest rates has been verified in many studies, as Edwards (1986), Alexander and Anker (1997), Lemmen and Goodhart (1999), Lønning (2000), Copeland and Jones (2001), and Codogno et al.

S. Zedda (✉)
University of Cagliari, via S. Ignazio, 74 09123 Cagliari, Italy
e-mail: szedda@unica.it

F. Amblard et al. (eds.), *Advances in Artificial Economics*, Lecture Notes in Economics and Mathematical Systems 676, DOI 10.1007/978-3-319-09578-3_11

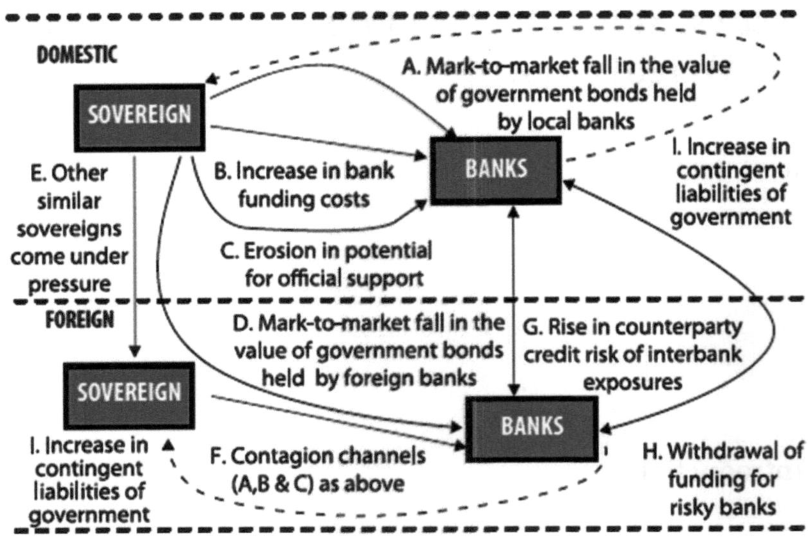

Fig. 1 Cross-relationship between sovereigns and banks. *Source*: IMF Global Financial Stability Report (2010)

(2003). With reference to the euro area, sovereign spreads are recognised as being mainly driven by debt, deficits, and debt-service ratios [see Bernoth et al. (2004), Bernoth and Erdogan (2010)].

Financial markets quickly internalise the effects described above: as observed in Acharya et al. (2011) contagion between the two sectors through the CDS market is really quick, differently from the translation of banking crises into public debt via higher deficit or via the real economy where contraction of economic activity is mainly due to reduced lending and augmented funding costs [see Cecchetti et al. (2009)].

The transmission of sovereign risk to banks riskiness and profitability was clear in recent years, as the increase in sovereign risk and the downgrade of several countries had a negative effect on bank riskiness. The direct impact of the recent sovereign debt crisis on banks' balance sheets was quantified in the European Banking Authority (EBA)'s 2011 technical proposal to the European Council. In its capital exercise, EBA was able to provide an overview of the sovereign portfolio of European banks, which represent the channel of transmission explicitly considered in our model.

About the linkages of the banking results to the real economy, while lots of studies have tested the influence of the banking activity on GDP, we only have a few papers that refer on how the GDP variations influence the banking results, loans riskiness and losses, e.g. Karimzadeh et al. (2013), Demirguc-Kunt and Huizinga (2000) and Bikker and Hu (2002) findings suggest that bank profits are correlated with the business cycle.

Athanasoglou et al. (2008), confirms that the business cycle significantly affects bank profits, even after controlling for the effect of other determinants strongly correlated with the cycle. Albertazzi and Gambacorta (2009) specify that "bank profits pro-cyclicality derives from the effect that the economic cycle exerts on net interest income (via lending activity) and loan loss provisions (via credit portfolio quality)". With reference to the interlinkages between bank and public finances, Galliani and Zedda (2014) analysed the cross relationship and the possible crisis worsening due to the circular nature of the relationship.

2 Methodology

Banks balance sheet can be represented as follows:

Assets	Liabilities and equity
Loans $\sum_k A_{ik}$	Equity K_i
Sovereign bonds $\sum_c SB_{ic}$	Deposits
Interbank credits $\sum_j IB_{ji}$ Other assets	Interbank debts $\sum_j IB_{ji}$ Other liabilities

As in the Basel II FIRB model, bank results can be represented as a one factor model, partially determined by diversified risks, but partially dependent on a non diversifiable component.

The diversified component is quantified as its expected value, so determined by the assets PD, while the undiversified component, linked to the macro variables, can be proxied by the GDP variation of the home country.

Banks are considered to be in distress as soon as losses become higher than capital:

$$L_i > K_i$$

The second source of losses is the interbank contagion, so that in case of distress of the bank j, the bank i, creditor for the amount IB_{ji} will experience a loss given by $IB_{ji} \times LGD_j$ where LGD_j is the interbank Loss Given Default (LGD) of the distressed bank j.

The third component of losses is due to the sovereign bonds market value reduction (haircut).

$$L_i' = L_i + \sum_j \left[IB_{ji} \times d(j) \times LGD_j \right]$$

2.1 Public Finances

The deficit variation is the result of the projected public budget result, $\overline{DEF_c}$, corrected for GDP variation multiplied by the sensitivity parameter of deficit to GDP for the country, S_c.

$$DEF_c = \overline{DEF_c} + \Delta GDP \times S_c$$

The sovereign bonds will then be affect by a variation in its market value. As a consequence, every bank i that invested in sovereign bonds of the country c will experience an additional loss. Summing up for all countries we have

$$L_i'' = L_i + \sum_j \left[IB_{ji} \times d(j) \times LGD_j \right] + \sum_c \Delta SB_{ic}$$

where $d(i)$ is the dummy variable set to 1 in case of default of the bank i and 0 otherwise. The second source of public finances instability we consider is the cost of bank rescuing.

Based on the previous paragraph representation, we have:

$$DEF_c' = DEF_c + \sum_i d(i) \times \left(L_i'' - K_i \right)$$

This additional public deficit will affect itself the banks stability, inducing more defaults, more contagion, more need to bank rescuing, etc.

2.1.1 Implementation

While some of the values are in banks balance sheet and can be easily be obtained from bank data as Bankscope, the distinction of sovereign bonds in each bank portfolio by issuing country is not reported, and is based on the EBA capital exercise data.

Another important issue refers to the bank assets riskiness, crucial for modelling the loan losses probability distribution.

Recently, Drehmann and Tarashev (2013) based their simulations on the a posteriori matching between the Moody's KMV estimates of the bank's PD and the one resulting from simulations. In the SYMBOL approach we use here (see De Lisa et al. 2011) the assets PD is obtained on the base of the minimum capital requirement, numerically inverting the Basel II FIRB formula:

The obtained $P\widehat{D}_i$ are then used to generate a set of correlated losses across all banks in the system. For each simulation s, calculate bank i's losses L_{is} performing a

Monte Carlo simulation based on the following representation of the FIRB formula:

$$L_{is}\left(z_{is}; P\widehat{D}_i\right) = \left[0.45 \times N\left[\sqrt{\frac{1}{1-R\left(P\widehat{D}_i,50\right)}}N^{-1}\left(P\widehat{D}_i\right) + \sqrt{\frac{R\left(P\widehat{D}_i,50\right)}{1-R\left(P\widehat{D}_i,50\right)}}N^{-1}\left(z_{is}\right)\right] - P\widehat{D}_i \times 0.45\right] \times \frac{1}{1-1.5\times B\left(P\widehat{D}_i\right)} \times 1.06$$

Where $N^{-1}(z_{is}) \sim N(0,1)\,\forall\, i, s$ is the random variable, representing the real economy business cycle possible results, that is correlated over banks so that $\mathrm{cov}(z_{is}, z_{is}) = 0.5\ \forall\, i \neq l$.

The simulation is typically stopped when 100,000 runs with at least one default are obtained.

The sovereign bonds haircut is estimated as follows:

$$\Delta SB_c\left(\Delta GDP_c\right) = \Delta GDP_c \times S_c \times YS \times YP_c$$

where S_c is the sensitivity of the country balance sheet to GDP variations, as estimated by the European Commission, DG ECFIN for surveillance purposes, *YS* is the yield sensitivity to deficit variations and YP_c is the parameter for converting the yield variation into value variation preserving the market equilibrium of expected returns.

Adding the Public Finances component, the first round contagion will be:

$$L''_{is} = L_{is} + \sum_j \left[\sum_i IB_{ji} \times \frac{\sum_j IB_{ij}}{\sum_i \sum_j IB_{ij}} d\left(j_s\right) \times 0.4\right] + \sum_c \Delta SB_c$$

3 Results

The preliminary model implementation is a simplified version, which does not include the weighted average of the countries and the real data on the correlation between GDP variations of the European countries.

Nevertheless the preliminary results show some interesting values.

In the following table the estimations for the "contagion" tail risk are reported, with different settings of the GDP variation correlation *R* from zero to 90 %, and including or not the exposures to sovereign bonds (see Tables 1 and 2).

Table 1 Crises dimension probability distribution without sovereign bonds exposures (billion euro)

Percentiles (%)	R = 0.9	R = 0.7	R = 0.5	R = 0
99.9	192.42	90.32	38.52	29.74
99.95	283.83	224.54	181.82	139.71
99.99	461.03	360.70	288.76	218.86

Table 2 Crises dimension probability distribution with sovereign bonds exposures (billion euro)

Percentiles (%)	R = 0.9	R = 0.7	R = 0.5	R = 0
99.9	337.01	279.31	59.51	34.20
99.95	414.48	363.52	313.46	255.04
99.99	565.50	484.46	420.22	330.39

Results show that not considering, or underestimating the correlation among countries, and not considering the effect of the circular relationship between banking systems and public finances leads to an important underestimation of the crisis possible effects. It is evident in all the three levels of percentiles, but in the first one, corresponding to the level set for the capital coverage in the Basel II regulation for single banks, it changes from 29.7 to 337, so more than 11 times more the "base" estimation.

References

Acharya VV, Drechsler I, Schnabl P (2011) A phyrrhic victory? Bank Bailouts and Sovereign Credit Risk, CEPR Discussion Paper 8679

Albertazzi U, Gambacorta L (2009) Bank profitability and the business cycle. J Financ Stability 5(4):393–409

Alexander V, Anker P (1997) Fiscal discipline and the question of convergence of national interest rates in the European Union. Open Econ Rev 8:335–352

Athanasoglou PP, Brissimis SN, Delis MD (2008) Bank-specific, industry-specific and macroeconomic determinants of bank profitability. J Int Financ Markets, Inst Money 18(2):121–136

Basel Committee on Banking Supervision (2010) The transmission channels between the financial and real sectors: a critical survey of the literature, Issue No. 27

Bernoth K, Erdogan B (2010) Sovereign bond yield spreads: a time-varying coefficient approach. DIW, mimeo

Bernoth K, von Hagen J, Schuknecht L (2004) Sovereign risk Premia in the European Government Bond Market. European Central Bank Working Paper Series No. 369

Bikker JA, Hu H (2002) Cyclical patterns in profits, provisioning and lending of banks. DNB Staff Reports, No. 86, Amsterdam

Cecchetti SG, Kohler M, Upper C (2009) Financial crises and economic activity, Working Paper, BIS

Codogno L, Favero CA, Missale A (2003) Yield spreads on EMU government bonds. Econ Policy 18(37):505–532

Copeland L, Jones S-A (2001) Default probabilities of European sovereign debt: market-based estimates. Appl Econ Lett 8(5):321–324

De Lisa R, Zedda S, Vallascas F, Campolongo F, Marchesi M (2011) Modelling deposit insurance scheme losses in a Basel 2 framework. J Financ Serv Res 40–3:123–141

Demirguc-Kunt A, Huizinga H (2000). Financial structure and bank profitability. Policy Research Working Paper Series 2430. The World Bank

Drehmann M, Tarashev N (2013) Measuring the systemic importance of interconnected banks. J Financ Intermed 22:586–607

Edwards S (1986) The pricing of bonds and bank loans in international markets: an empirical analysis of developing countries' foreign borrowing. NBER Working Paper No. 1689

European Commission (2011) Directorate General for Economic and Financial Affairs, Public finances in EMU 2011. European Economy 3, 2011. Available at http://ec.europa.eu/economy_finance/publications/european_economy/2011/ee3_en.htm

Galliani C, Zedda S (2014) Will the bail-in break the vicious circle between banks and their sovereign? Comput Econ. doi:10.1007/s10614-014-9436-9

International Monetary Fund (2010) Global financial stability report. Sovereigns, Funding and Systemic Liquidity. World Economic and Financial Surveys, October, 2010. Available at http://www.imf.org/External/Pubs/FT/GFSR/2010/02/index.htm

Karimzadeh M, Akhtar SMJ, Karimzadeh B (2013) Determinants of profitability of banking sector in India. Transit Stud Rev 20(2):211–219

Lemmen J, Goodhart C (1999) Credit risk and European government bond markets: a panel data econometric analysis. Eastern Econ J 25:1

Lønning I (2000) Default Premia on European Government Debt. Weltwirtsch Arch 136(2): 259–283

Saudis and Expats: An Agent-Based Model of the Saudi Arabian Labor Market

Davoud Taghawi-Nejad

Saudi Arabia like the other states of the Arabian Peninsula faces an unprecedented challenge: How to transform an economy that was largely driven by foreign labor, that was bought with oil money, into an economy that employs its own people and that creates wealth without depleting its oil wealth. The challenge to 'Sauditize' the economy cannot be guided by past experiences in other countries, simply because there are none.

We are developing an agent-based decision support system that enables Saudi policy makers to explore different policy options: such as hiring quotas for Saudis as well as taxation of expat labor and minimum wages for expats or expats and Saudis alike.

This agent-based labor market model specifically captures the particularities of the Saudi Arabian labor market. The Saudi Arabian labor market is special as it is characterized by a very low participation of Saudis in the private sector labor market, about 680,000 out of 20 million inhabitants. The low participation rate is caused by firms preferring not to hire Saudis as well as by Saudis choosing not to work in the private sector.

Foreign workers might be more attractive to firms because they have lower bargaining power. Out of the ca. seven million expatriates, 6.5 are low and medium-skilled and come from very poor countries (Hertog 2010). Therefore they are willing to work for very low wages and put a great deal of effort into their work. What is more expatriate workers come on sponsored visas. That means that under the threat of deportation they can only work for their sponsoring firm. Thus once they are in Saudi Arabia their employer can act as a quasi-monopolist, this additional factor forces them to put in much more effort than Saudis and accept lower wages. What is more there are disincentives for Saudis to work: temporary

D. Taghawi-Nejad (✉)
ESD University, Massachusetts Institute of Technology, Cambridge, USA
e-mail: davoud@taghawi-nejad.de

F. Amblard et al. (eds.), *Advances in Artificial Economics*, Lecture Notes in Economics and Mathematical Systems 676, DOI 10.1007/978-3-319-09578-3_12

unemployment assistance and high public sector employment. Job security and a lax work environment mean the government is still the preferred option (Hertog 2010).

Saudi Arabia implemented the Nitaqat quota system to increase the participation of Saudis in the private labor market. Nitaqat establishes minimum quotas for Saudis in each sector. The quotas vary with firm size. Firms that employ a lower percentage of Saudis do not get new visas issued for expatriate workers.

The Nitaqat quota system poses a particular challenge on the model. Firms can fulfill the quota by firing expatriates or hiring Saudis. When firms choose to expand they can hire a combination of Saudis and expatriates. The combination of foreign and Saudi workers now becomes subject to strategic hiring. In order to capture this strategic behavior in the model, firms hire a combination of foreign workers and Saudis, on the basis of a discrete optimization algorithm. It has even been observed that firms hire unproductive Saudis, who might even not be expected to show up to work, in order to hire productive expatriates (Hertog 2010; Sehgal 2013) While theoretically the model can capture an offer of pretending to work for a wage, it is currently not employed as we are lacking the data to calibrate the model accordingly.

1 Policy Questions

The objective of the model is to analyze policy recommendations for their potential to create employment and build up the country's capability to create wealth, without overly relying on expatriate labor or oil. The core policy to be analyzed is a sector-specific quota system that established minimum percentages for domestic workers for firms. A series of alternative policies are also implemented: the introduction of sector-specific minimum wages, both for expatriates only or for all workers in the labor market; taxation on labor in general; creation of an internal market for expats; changes in the length of visa, probation period and quasi-tenure after 3 years.

With regards to these policies the most obvious question is how quotas impact on the number of employed Saudis. But a mere headcount falls short of reflecting the real situation. It is important to ask how Nitaqat or hypothetical variations of Nitaqat impact on the wage distribution and the creation of wealth/GDP. An extension not discussed in this paper also analyzes the creation of fake employment to fulfill quotas and a Saudi low wage sector.

2 Empirical Findings

Between November 2011 and February 2012, Saudi Arabia faced a comprehensive quota system—Nitaqat—for firms in all sectors. Jennifer Peck in her paper: "Can Hiring Quotas Work? The Effect of the Nitaqat Program on the Saudi Private Sector", (Peck 2014) finds that the during the program Saudi employment increased by 96,000, but only 73,000 can be attributed to the program, the rest would have been created without it. Further, Nitaqat had significant negative effects on employment. Overall private sector employment decreased by 418,000 workers.

One particular phenomenon observed is that of poaching. Firms that were not in compliance with the Sauditization quota, hired away Saudis from compliant firms. While the quota could be met by both decreasing the number of expatriates as well as increasing the number of Saudis, firms largely increased Saudi employment. At this point, we cannot distinguish between real and fake employment. But research on this topic is underway.

As an example in this paper we will show and simulate the manufacturing sector.

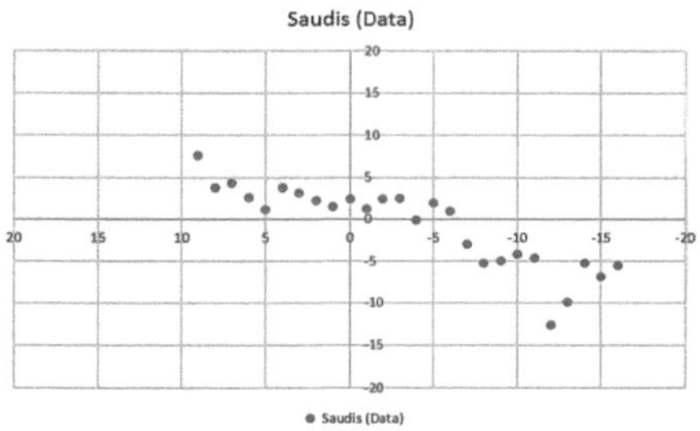

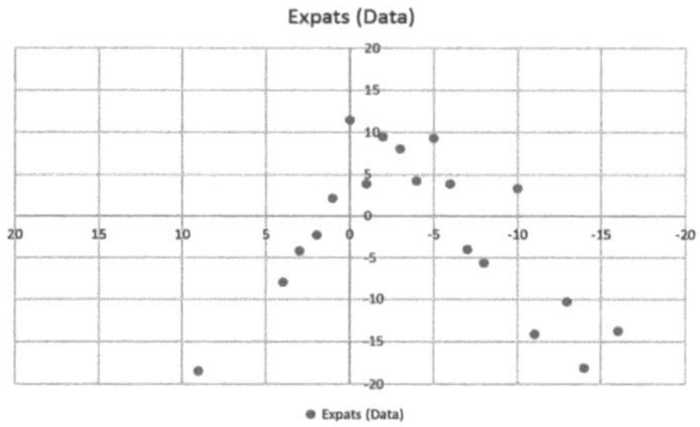

In the two graphs above, every point on the x-axis represents the firms that needs to hire a certain number of Saudis to comply with Nitaqat. For example the group of firms in the at x = 5, for example need to hire 5 Saudis to comply with the quotas. Firms with a negative x value could lose Saudis and still be compliant. The y-axis on the first graph, shows the average number of Saudis firms in this group actually hired by October 2012. The second graph has the same x-axis but shows the number of expatriates the firms had to lay off. The x-axis is restricted to values we can reproduce with a 1–100 scale model of the manufacturing sector.

We can observe that firms increased Saudi employment rather than decrease expatriate employment. On the first graph we see that non-compliant firms on the left hire Saudis, but overly compliant firms lose Saudis. The phenomenon of hiring away Saudis from compliant firms is called poaching, first described in Jennifer Peck's paper. The expatriate graph has an inverse U-shaped response, strongly non-compliant firms were forced to decrease the number of expatriates, but also some over-compliant firms decreased the number of expatriates.

Time-series data will be discussed along with the modeling results.

3 Model

Accessing the reactions of the Saudi labor market to policies requires a model that captures the behavioral responses of workers as well as interaction of workers and strategically-behaving firms. Modeling the behavioral responses of workers would be meaningless without capturing the heterogeneity between different types of workers in the Saudi labor market. To address all these issues the model has five components: workers, firms and two newspapers for Saudis and Expats and final goods markets.

Workers are either Saudi or expatriate, they have a corresponding reservation wage derived from empirical data. Unemployed workers apply for a job and when hired provide labor. Saudi workers can also engage in on-the-job searches.

Firms use labor as their only source of production. They set prices, wages and a target production capacity according to their observed demand and labor supply. Firms hire new staff when the current production capacity is below the target production capacity and the individual hire is profitable. If minimum wages or quotas are in place, firms comply with them. When quotas are in place, firms hire Saudis even when the individual hire is unprofitable but allows the firm to hire productive expats. Firms then quote their price and produce the demanded amount of goods, if their production capacity suffices. Profits are distributed.

There are 52 sectors, each company produces for one sector. The demand in each sector is derived and calibrated based on the Saudi supply and use tables.

3.1 Overview

Firms and workers are individual software agents. Every round, which represents a day, they go through a seven-step process:

1. Firms advertise for Saudi and expatriate labor, quoting their respective wages in a newspaper.
2. Workers apply, when their reservation wage is met.
3. Firms hire/fire (according to visa-status and tenure).
4. Firms quote a price and observe the demand. If their production suffices they sell the quantity demanded at the quoted price.
5. Firms pay wages and distribute profits.
6. Firms adjust their prices and wages for new hires.
7. Bankrupt firms get removed.

3.2 Modeling of the Workers

The workers in this model have individual characteristics: visa status; Saudi, non-Saudi a reservation wage and individual productivity. These attributes are derived from data from the Ministry of Labor and from the social insurance entity GOSI. In a future extension of this model, the individual characteristics of workers will be modeled in more detail. The reservation wage will then be a function of the workers' characteristics.

Workers in this model are not consumers. The final goods market is fixed and its parameters are part of the calibration. In the context of Saudi Arabia that is much less of a restriction than it would be for other countries, because the demand is primarily driven by oil-financed government expenditures and therefore to a large degree independent from the workers' income. What is more the large share of remittances in the labor market that have a low share in the current account balance also decouple the income from sector demand.

The two secondary workers' characteristics that impact the model directly are their reservation wage and their productivity. Reservation wage and productivity are sector-specific. When workers are generated these characteristics can be conditional on primary characteristics such as age, education, Saudi/non-Saudi, family income... Currently the distribution and mean is unconditionally calibrated.

A worker is either employed or unemployed. Saudis and expats apply at one firm per day, if they find a firm that offers a wage that meets their reservation wage. Saudis also engage in the job search. Their search intensity is a parameter, which we use for calibration.

3.3 Firms

A firm produces a representative good of one of the 52 sectors. Goods in each sector are differentiated. Firms use labor as their only source of production using a constant economics of scale production function.

3.3.1 Price, and Production Target Setting

Firms planned production and prices are set adaptively. Planned production is adaptively increased when there is excess demand—decreased in the opposite case. Prices are modified whenever the firm doesn't find workers to meet its production target (which feeds back to the observed demand). The price-setting mechanism in the final goods market is inspired by (Gaffeo et al. 2008).

A firm's production is:

$$x_{i,t} = \sum_j a_{i,j}$$

where a is the productivity of worker j in firm i. A firm's production target (planned production) is $pp_{i,t}$.

When observed demand exceeds/falls short of planned production, planned production is increased/decreased. Planned production falls never below the actual demand and is unaltered when it is close to actual production.

$$d_{i,t-1} < pp_{i,t-1} - \overline{a}_{i,t} \quad pp_t = \max\left(d_{t-i}, pp_{t-1}\left(1 - \sigma_\tau\right)\right)$$

$$d_{i,t-1} > pp_{i,t-1} \quad pp_t = \min\left(d_{t-i}, pp_{t-1}\left(1 + \sigma_\tau\right)\right)$$

When a firm's production exceeds their planned production by more than the productivity of the average worker in that firm, the firm decreases its price. But never below its marginal costs. If the production falls short by more than the average productivity of a worker, the price is increased. The increase/decrease of the price is a uniform random percentage. The mean is a sector parameter.

$$x_{i,t} > pp_{i,t} + \overline{a}_{i,t} \quad p_{i,t+1} = \max\left(p_{i,t}\left(1 - \sigma_\eta\right), \overline{a}_{i,t}/\overline{w}_{i,t}\right)$$
$$x_{i,t} < pp_{i,t} - \overline{a}_{i,t} \quad p_{i,t+1} = p_{i,t}\left(1 + \sigma_\eta\right)$$

3.3.2 Wage Setting

The wage a firm offers to new workers is determined separately for Saudis and expatriates. The respective advertised wage is the firm's average wage in the respective category plus a random Gaussian variable. The standard deviation is a model parameter.

$$w^{o,s}{}_{i,t+1} = \overline{w}_{i,t}{}^{s}{}_{i,t}\left(1 + \sigma_\phi\right)$$
$$w^{o,e}{}_{i,t+1} = \overline{w}_{i,t}\left(1 + \sigma_\phi\right)$$

While a wage offer is a random variation, the wages actually accepted by the worker and the firm, by hiring the worker at this wage is subjected to market forces.

Firms adjust wages if minimum wages are required by law and add taxes[1] if applicable.

3.3.3 Firms Hire and Fire

When a firm has a production target above the current production they hire new workers. All hiring decisions are profit-maximizing given the current prices and assuming that the current planned production is the maximum the firm can sell. Firms can observe workers' productivity. A firm keeps all workers which cannot be currently laid off. From the stack of workers with expired contracts or visas and new applicants a firm employs the most profitable subset that is in accordance with the quotas and other laws.[2] The profitability is calculated assuming the current prices

[1]Currently Saudi has no taxes on labor.

[2]It is also possible to model partial compliance.

and that only the planned production can be sold. Firms cannot hire more workers than it can pay from their net worth. Employees that are not kept are laid off.

When the planned production is below the current production workers are laid off, if it's possible and profitable to do so.

Firms can only fire Saudis in a 90-day period and at contract end. After 3 years Saudis receive tenure and cannot be fired. We assume 1-year contracts before this date. Expatriate labor can be fired at the visa expiration: 1 year. Firms can not renew visas for expatriates if a quota is binding.

3.3.4 Selling, Paying Wages and Distributing Profits

Firms quote the price to the (representative agent) final good market, which in turn determines the demand $x_{i,j}$ for their good. Firms sell the good and receive their revenue $p_i x_i$. Workers are paid their wages. If a company's profits are lower than a certain percentage of their net worth, profits are distributed.

3.3.5 Bankruptcy

Bankrupt firms are removed from the simulation.

4 Final Goods Markets

There are 52 sectors, with an endogenous number of firms in each sector. The firms create differentiated goods. As this model is interested in the labor market, we use a standard representative agent framework to model the final goods markets. The demand for each sector is represented by a Cobb-Douglas utility function, where the demand an individual firm encounters is determined by a CES utility function to represent diversified products or geographical diversity of the firms. The final goods markets are calibrated to the input–output tables and are represented by a Dixit-Stiglitz type monopolistic competition (Dixit and Stiglitz 1977). $u = U(V_1(x_1, x_2, \ldots, x_n), V_2(.), \ldots, V_n(.))$ where U(.) is Cobb-Douglas and V(.) is a constant elasticity of substitution function. This leads to the following demand equations:

$$d_{i,j} = \frac{\alpha_i I}{q_i}\left[\frac{q_i}{p_j}\right]^{1/(1-l)} \quad \text{where} \quad q_{i,j} = \sum\nolimits_j^n p^{l\backslash(l-1)}$$

where i is the index of the firm and j the sector. I is a modeling parameter.

For the single sector simulation exposed in the remainder of this paper, we will only use the inner CES function. The two parameters to calibrate it are love of variety and the share of income spend in this sector.

5 Calibration

5.1 Strategy

We calibrate the model using a Kriging model:

The agent-based model is run with different parameters. The parameters are chosen employing a Latin hypercube sample technique, to insure optimal efficiency. Runs are repeated with different random seeds. The outcomes of the simulation are transformed with a weighted sum of squares to reproduce certain stylized facts. The correspondence between the parameters and the weighted sum of squares is employed to build a Kriging model. The parameter with the lowest sum of weighted squares is our best candidate parameterization (BCP). A Kriging model employs a Gaussian process to build a meta model $f(x_1, x_2, \ldots, x_n) = ss$. The Kriging model asymptotically produces the same results as the agent-based model, but is several hundred times faster. We use the sweep of the Kriging model to find a small number of candidates for a parameterization that reproduces the stylized facts best. In other words we run the simulation for every point on a multi-dimensional lattice that spans all parameters. We double check with the real model that this candidate is indeed better than the best candidate parameterization (BCP). If we find a new best candidate we narrow down our search space and make this candidate our new center of the search. If no better new candidate is found, but the best n candidates did correctly predict the agent-based model's weighted sum of squares, we also narrow down the search space. After this, we run the agent-based model again, restricting the parameters employed in the Latin hypercube to the smaller search space and the process continues from the beginning.

5.2 Calibration Criteria

The simulation is a scale model of reality, we simulate for example one firm and one worker for every 100 firms or workers in the economy.

We minimize the sum of squares of a set of stylized facts at two points in time to correspond to our dataset:

- Number of expatriates employed in each sector
- Number of Saudis employed in each sector
- Output in this sector
- Average Saudi wages in each sector
- Average expatriate wages in each sector

The parameters we systematically vary using a Latin hypercube sampling technique are:

- Love for variety in the demand function for a single market
- The number of expatriates that are in the labor pool for each market
- Productivity mean and variance of expats (later, that should be a conditional mean and variance)
- Productivity mean and variance of Saudis
- Reapplication probability
- Mean and variance of the reservation wage of expats
- Mean and variance of the reservation wage of Saudis
- Sector spending

On the other hand, the parameters we obtain from our data and the policies are:

- Number of Saudis in the labor pool
- The number of firms
- Initial policies
  - Expat minimum wage
  - Expat tax per head
  - Expat tax percentage
  - Saudi minimum wage
  - Saudi tax per head
  - Saudi tax percentage
  - Saudization percentage
  - Visa length

5.3 *Validity Check*

The simulation needs to reproduce the change in the stylized facts for at least one policy change, using data after the day of the policy change.

Below we show the number of Saudis hired for a particular compliance level and in the second graph the number of expatriates lost accordingly. We see that the model reproduces observed reality in the manufacturing sector.

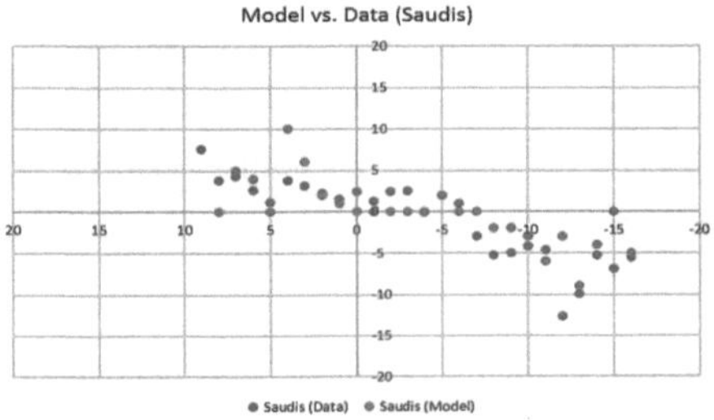

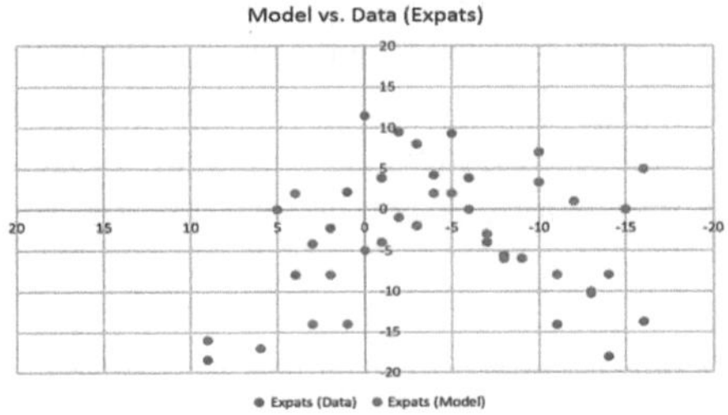

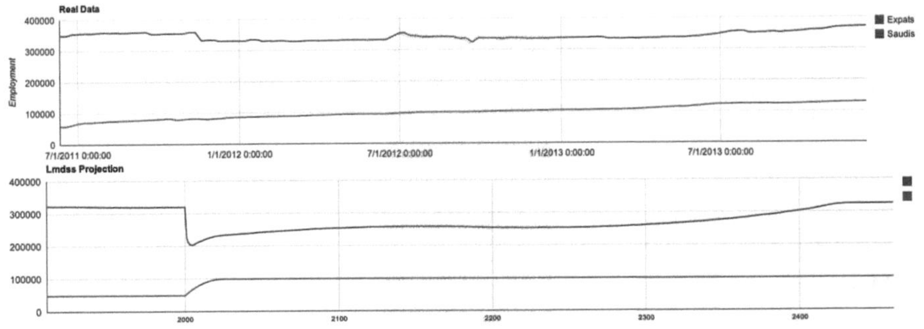

The timeline of the real data and the simulation show that our simulation captures the essential dynamics of the policy change, although our levels are not yet completely correct. As expected the model exhibits less fluctuations. While we can observe preparatory hiring of Saudis in the real date, the inability to think about the future makes the software agents incapable of this behavior.

6 Policy Simulation

There is a large debate whether a quota system is the best response to the low share of Saudis in the private labor market. Instead of a quota system we introduce a mandatory minimum wage for expatriates of 1,000 riyals every month. We create an as-if scenario pretending the minimum wage would have been introduced instead of a quota system.

We can see that even with myopic agents, the transition is smoother and the effect is sustained for a longer time.

Conclusion
Employing agent-based modeling and Kriging to calibrate the model, we created a model of the manufacturing sector of Saudi Arabia. Other sectors will follow. We are able to reproduce a policy intervention: Nitaqat. The effect of its quotas is captured in the model; both in the created time-series as well as in a closer inspection of cross-sectional data on the firms' responses sorted by compliance level. The model can be used to explore a variety of alternative histories, simulating alternative policies, such as expatriate taxes and minimum wages for expats or for expats and Saudis. In the same manner, future policies can be tested before they are implemented.

Acknowledgments This project was undertaken as part of the Labor Market Decision Support System at the Center for Complex Engineering Systems at MIT and the King Abdulaziz City for Science and Technology. I thank Jennifer Peck, whose empirical work was the basis of this policy simulator and for her support and discussions. As well as the CCES and the LMDSS team for their work.

References

Dixit AK, Stiglitz JE (1977) Monopolistic competition and optimum product diversity. Am Econ Rev 67(3):297–308

Gaffeo E et al (2008) Adaptive microfoundations for emergent macroeconomics. Eastern Econ J 34(4):441–463

Hertog S (2010) Princes, brokers, and bureaucrats: oil and the state in Saudi Arabia. Cornell University Press, Ithaca

Peck J (2014) Can Hiring Quotas Work? The Effect of the Nitaqat Program on the Saudi Private Sector, Working Paper, KACST and MIT March 2014

Sehgal R (2013) Nitaqat law: will it solve Saudi Arabia's unemployment problems? – analysis. Eurasia Review 2013. http://www.eurasiareview.com/10072013-nitaqat-law-will-it-solve-saudi-arabias-unempl[oyment-problems-analysis/

Forbidding Fixed Duration Contracts: Unfolding the Opposing Consequences with a Multi-Agent Model of the French Labor Market

Olivier Goudet, Jean-Daniel Kant, and Gérard Ballot

1 Introduction

The model WorkSim is a new tool of analysis for the French labor market. The first objective of the model is to reproduce the gross flows between six important states for the workers namely student, employment—distinguishing between fixed duration contracts (FDC) and open ended contracts (OEC), unemployment, inactivity, and retirement. The two main novelties of the model are that it simulates the gross flows of workers between the six states on the basis of the rational decisions of individual heterogeneous agents, and that both firms and individuals, the latter linked to households, are represented. Once calibrated by an optimization algorithm, in order to obtain a steady state situation for a large number of aggregate targets corresponding to 2011 data, the model is a tool for experimenting labor market policies, including changes in the labor law. In the present paper, we focus on experiments dealing with the suppression of the FDC, which have an important place in the French labour market. Even though the FDC represent in 2011 only 10.1 % of the workers employed in the private sector against 84 % for the OEC, 3.6 % for the temporary help workers, and 2.2 % for the apprentices, they count for 80 % of the hires. Moreover the number of short FDC has increased steadily since 2003. The number of FDC of 1 week or less has increased by 120 % over the period 2000–2010, and the number of FDC between 1 week and 1 month has increased by 36 % while the number of FDC of more than 1 month has remained almost stable

O. Goudet (✉) • J.-D. Kant
Sorbonne Universités, UPMC Univ Paris 06, UMR 7606, LIP6, 75005 Paris, France

CNRS, UMR 7606, LIP6, 75005 Paris, France
e-mail: olivier.goudet@gmail.com

G. Ballot
CRED EA-7321, Université Paris 2, TEPP FR CNRS 3435, Paris, France

F. Amblard et al. (eds.), *Advances in Artificial Economics*, Lecture Notes in Economics and Mathematical Systems 676, DOI 10.1007/978-3-319-09578-3_13

with a 2.5 % increase (Berche et al. 2011). The same source points to a decrease of the OEC of 6.6 % .The high and increasing turnover of the FDC is likely to have important effects on the workings of the labor market, and the unemployment experience of the workers who go through these FDC. There is presently a strong controversy about the *effects of FDC on unemployment*.

Buffer stock effect—Employers consider that they are a crucial necessity to respond to short term fluctuations of demand since the cost of firing workers on OEC is high (see for instance the study by De Froment 2012). Then the interdiction of FDC would, when an employer faces an uncertain demand, induce him to hire less than he would have if allowed to recruit on FDC, when demand rises, and steady state unemployment would be higher. The simple argument stating that higher firing costs lower employment in market with only OEC has however been questioned by several economists since high firing costs also lower dismissals, and the volatility of demand in this context of costly adjustment is another important determinant of employment. As Bentolila and Bertola (1990) have shown in an influential paper, the issue is complex, the net theoretical effect on steady-state unemployment is ambiguous, and simulations suggest that unemployment could be very little affected.

Screening effect—Other economists (Faccini 2008; Bucher 2010; Berson and Ferrari 2013) have pointed out that FDC are used to screen young workers (with no firing cost) before giving an OEC to the most able ones.

Experience effect—Ballot has a somewhat different argument but in the same line Ballot (1981, 2002): the FDC offer some increase in experience which raises the probability to access an OEC, either in the same firm or also in another firm.[1]

Stepping stone effect—The two last effects converge to say that if FDC did not exist, the firms would be reluctant to hire those of the inexperienced workers who do not have recognized diplomas, especially as they have little information on their productivity. The direct recruitment of these young workers would be lower, hence steady state unemployment higher. This convergent impact allows us to group the *screening* and the *Experience* effects under the common label of a *stepping stone effect*.

Churning effect—Other economists such as Blanchard and Landier point out that the termination of the FDC, given the interdiction to renew indefinitely (in France more than once such a contract), sends the workers to unemployment frequently (Blanchard and Landier 2002). The employers are assumed to be very reluctant to give an OEC to the workers at the end of the FDC, given the higher expected cost of OEC. This is called the *churning phenomenon* and it increases unemployment. Other motives for the use of FDC such as replacing ill workers exist also. *No analytical model presently takes into account all these effects which go in opposite directions*, for reasons of complexity, and economists presently do not know if the net effect of FDC on unemployment is positive or negative. Finally natural experiments are not available since no government has forbidden FDC, and

[1]For some empirical evidence see Booth et al. (2002).

there is no neat reverse experiment in which a government has suddenly allowed FDC.[2]

Our agent based model is able to include the role of FDC as a screening device with an *experience effect*, and it also accounts for the important flows out of FDC into unemployment (the *churning effect*). In a nutshell, The WorSim model is a two sided search model of the labor market, with individuals searching for a job, either as unemployed or on-the-job, if they are not satisfied with the total utility provided by the job, wage and conditions of work. They have a reservation utility. They may also give up searching since it has a disutility in terms of time, become inactive and rely on welfare or family income. Firms post vacant jobs with a wage, and set a minimum productivity for hiring a candidate. If candidates apply, the firm evaluates their productivity with a noise, and selects the best, or none if the best does not attain the minimum productivity. This modeling strategy leads to a search model with microfoundations which respects the heterogeneity intrinsic to the search approach in economics. It strongly differs with the matching models in the line of Mortensen and Pissarides (1994) which assume the existence of an aggregate matching function. WorkSim therefore allows for non linearities in the hiring of competing heterogeneous workers, non linearities which are likely to be central in the workings of a labor market with two types of jobs. WorkSim has three additional features that make this model differ from matching models. Decisions are based on bounded rationality, there are keynesian features (downward rigidity of nominal wages of insiders, and a minimum wage, but new hires wages are influenced by the tension on the labor market), and lastly true productivity of a worker is learned by the employing firm over time but never perfectly known. Most of these features are essential to determine the gross flows, and do realistic policy experiments.

2 Model Outline

In this section, we outline the main features of the WorkSim Model.[3] There are three types of *agents* in WorkSim: Individuals, Firms and a Public Sector that recruits employees, collects payroll taxes on businesses, and sets (exogenously) public policies for the Labor Market. In addition, the model uses three *artefacts* (in the sense of Omicini et al. 2008) to deliver essential services to the agents:

- *JobAds*, which receives job offers from the firms and job applications from the job seekers. Dissemination of information, however, is based on the job search process, according to the principles of the theory of search Phelps (1970).

[2]Boeri in his survey of the literature on employment protection does not mention such an experiment (Boeri 2011).

[3]A more detailed presentation of this version of WorkSim could be found in our reference paper here: [http://www-poleia.lip6.fr/~kant/WS1.6_MCR.pdf]

- a *demographic* module, which manages the demographic processes in the model (retirement, death . . .).
- a "*statistical institute*" that calculates some statistics needed by the agents (e.g. tension in labor market), and disseminates information to a limited number of agents.

One distinctive feature of the WorkSim model is to integrate a fairly complete and flexible institutional framework that includes (1) the necessary elements of the French labor Law (including two types of contract: *fixed duration contracts* (*FDC*) and *open ended contracts* (*OEC*), permanent layoffs and discharges, redundancy payments, . . .), and (2) government decisions (minimum wages, welfare benefits, . . .).

The *simulation cycle* includes four main steps, as shown in Fig. 1 below:

1. Firm decisions: manage contracts & vacancies, evaluations, job creation / destruction;
2. Individual decisions: entering/leaving labor market, job search;
3. Firm decisions: manage applications and promotions;
4. Demography: household dynamics, retirements, aging.

The length of one period (i.e. one tick) in the simulation cycle corresponds to *1 week* in the real world, in order to take into account very short term contracts (1 week duration) that are found in the French labor market. Moreover, when statistics are needed, we took 2011 as a reference year, when we could find the most recent and complete statistical data and sources.

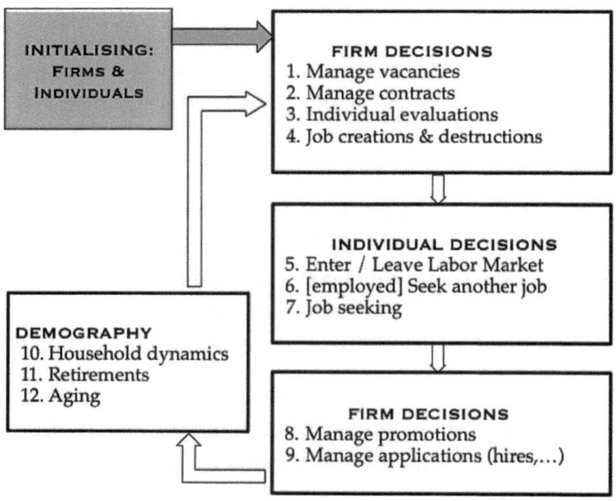

Fig. 1 The simulation cycle in WorkSim

2.1 Firm Decisions

In the current version of WorkSim, there is one good, and each employee produces a certain amount of a variety of this good which is unique to the firm but different from other firms variety only by the heterogeneous preferences of the consumers for this variety, and the fluctuations of these preferences (horizontal differentiation). The price is then unique and fixed. The only production factor is the labor, like in many agent-based model of the labor market. Therefore, the firm production is linear additive in terms of hours of work, and some employees only work part-time.

Productivity and salary—Positions and employees are not identical in terms of productivity. There is a base productivity attached to each position, and the employee will modulate its value, through his general productive characteristics (core productivity and general human capital built from work experience), and the specific human capital in the position he holds. Moreover, the employer has only an imperfect and evolving information on individual productivities. A hiring salary is also set for each potential position. It includes the minimum wage in France (SMIC), the base production modulated by the margin taken by the firm and the human capital of the employee.

Firm demand—At start, employees are randomly assigned to each company, which determines each global productions companies $Q_{j,t=0}$, with $j = 1..N$ and N is the total number of firms. We assume initially that the requested demand $D_{j,t=0}$ to one particular firm corresponds to the quantity produced plus an additional initial share representing a margin that will allow it to create its first job vacancies.

The global demand for all the firms in our model at $t = 0$ is $D_{t=0}^{total} = \sum_{j=1}^{N} D_{j,t=0}$.

In our economy, the market share of the firm j at $t = 0$ is given by $MS_{j,t=0} = \frac{D_{j,t=0}}{D_{t=0}^{total}}$.

We assume that the distribution of this demand varies between firms. Therefore we apply a stochastic shock on the market share of each firm each period (random walk):

$$\forall t,\ MS_{j,t} = MS_{j,t-1} \times (1 + \mathcal{N}(0, \sigma_{MS})) \tag{1}$$

σ_{MS} is an exogenous parameter.

Therefore the demand of the firm j is updated each period with this new market share:

$$D_{j,t} = \frac{MS_{j,t}}{\sum_{k=1}^{N} MS_{k,t}} \times D_t^{total} \tag{2}$$

Job creation—During job creation process, job characteristics depend firstly on the choice between FDC and OEC: this choice is made randomly according to the same probabilities for all firms, which differ depending on the qualification. Three levels of qualification are currently considered in WorkSim: employee or worker, middle level and executive. If a FDC is drawn, its duration will be determined again by drawing, according to qualification. Then, for the position to be definitively created, the firm estimates its expected profit and accounts for:

- the expected productivity of the position estimated by prospection (with informations about several randomly drawn job seekers)
- the expected salary of the position with payroll taxes the firm has to pay for each employee
- the expected total cost of vacancy amortized over the expected duration of the contract (cost of publication of the job, time devoted to interviews by recruiters…). This cost is estimated by learning.
- the end cost of the contract: severance pay for FDC contract (10 % of the cumulated wage paid during the contract) or potential firing cost for OEC contract estimated by learning.

If the profit is positive, a vacancy is created and sent to a limited number of job seekers per period through JobAds.

Vacancy destruction—In case there is a significant reduction in its demand, the company plans to remove one or more vacancies. Moreover, the vacancies that are not filled in excess of a period of higher vacancy than a certain threshold are deleted.

Hiring—A two-step process, the process of sorting symmetrical positions by job seekers, takes place:

(a) Firstly the firm assigns a score to each candidate (internal or external), this score is the expected profit for the company in the event the candidate is hired; then the best (highest profit) candidate is selected.
(b) Thereafter, the firm verifies that the candidate passes the *hiring norm* (computed by the firm as a profitability threshold, taking also the labor market tension into account). If this is the case, the candidate is hired; otherwise, the position remains vacant.

Employee management—Employee evaluation takes place: (1) at the end of the trial period for FDC and OEC, (2) at the end of FDC contract to decide if it should be renewed, (3) at the end of FDC contract, if the transformation of FDC to OEC is to be considered and (4) every year, at the contract's anniversary date, for each FDC employee. In order to decide whether the employee should be kept, the firm calculates a profit for each scenario (to keep the employee or not). Moreover, at each period, each employee has a fixed probability to be fired for a professional misconduct "observed" by the employer.

Economic firings—An evaluation of the financial viability of the company is done every month. The first date of the balance sheet is drawn randomly, then balances take place every month from this date. The company calculates annual profit made and monthly costs: if its ratio is below a certain threshold, the company has a loss and starts an economic firing process: (1) all vacancies are removed, and (2) the company fires "randomly" a number of employees on economic ground. This number is chosen as the minimum number of persons to fire in order to return above the profit threshold.

Firm bankruptcy —If a company is reduced to its managing director and is not profitable anymore, so this director has to be fired, the firm is considered bankrupt and the director becomes unemployed. However, we decide in our simulations to keep the number of firms constant. Hence, when a bankruptcy has occurred, we randomly select an active individual in the simulation who creates a new firm and sets him as a double role of managing director (in OEC) and producer.

2.2 Individual Decisions

The individuals take decisions at each period of the simulation. This decision process is modeled with a *state machine*, where one individual will be in one particular *state*: student, inactive, unemployed, employed not searching, employed searching a new job or retired. The transitions between these states can be caused by individual choices (for example: to start studying, to quit a job...), by external events (firing, death...), or eventually by a sequence of two decisions (applying for a job, and the firm hires the candidate). The decision-making process of individuals is sequential: the transition from one state to another is done by comparing the utility level of the current state with the expected utility level in a new state.

The utility functions have the generic form of a Cobb-Douglas function:

$$U = (Income + Amenity + Stability)^{1-\alpha}(Free\ Time)^{\alpha} \tag{3}$$

This utility function is a weighted aggregation of four factors:

Income: weekly income in euros (salary per week in employed state, unemployment benefit when in unemployment state, ...)

Amenity: non-monetary features perceived by the individual (social recognition, working environment, job hardness...), converted into a percentage of salary, and expressed in euros

Stability: criteria reflecting the preference of the individual for stability, i.e. for a job with the longest possible remaining contract duration. The maximum value is given for a permanent position (OEC) because its duration is—theoretically—indefinite. This stability is converted here into a percentage of salary and is expressed in euros.

Free time: free time per week available for the individual outside his working and search hours.

The parameter $\alpha \in [0, 1]$ encodes the preference of the individual for leisure or work.

Job search—The job search is modeled as a four steps process and only concerns agents in unemployed state or employed and searching for a new job:

1. At start, the individual requests from *JobAds* a list of vacancies matching his qualification and also some with an upper level of qualification.

2. He applies to the best offer he receives that exceeds his reservation utility
3. If the job seeker does not receive offers that match his level of qualification or all of his applications are rejected, he lowers his reservation utility.
4. Employees (simply employed or on-the-job-search) whose seniority in the firm exceeds a threshold parameter apply automatically to internal offers whose qualification is strictly a level above their own level.

3 Model Scaling Method

First of all, we scale the number of firms of the private sector (source INSEE 2011a). The factor retained is 1/4700 for the number of firms and individuals. With the total number of jobs obtained during this firms scaling, we adjust the initial number of "employed" individuals in the model. In total, we obtain 808 firms with 4,411 individuals in these private firms. We add public servants in a proportion of 21.3 % (source INSEE 2013b) to this number of employees in the private sector. Next we include the numbers of "inactive", "unemployed", "retired" and "student" agents corresponding to 2011 statistics (INSEE 2011d). In the end we obtain 8,713 individual agents and 808 firm agents, for a total of 9,521 agents in the simulation.

4 Model Calibration

4.1 Minimization of a Fitness Function

To calibrate the model parameters (35) we minimize a *fitness* function which is the weighted sum or the relative spreads between the outputs of our model and real targets of the French labor market in 2011 (58 targets overall) regrouped in different categories:

- Seven targets on unemployment rate by age range and by qualification level (source INSEE 2011c)
- Six targets on activity rate by age range and by gender (source INSEE 2011b)
- Twenty one targets on wages by age range and by qualification levels, and annual wages distribution per decile on the global population (source INSEE 2013a)
- Nine targets on labor flows (source DDMO/DARES 2011)
- Nine targets on annual transition rate (source employment survey *Enquête Emploi* (Jauneau and Nouel de Buzonniere 2011)).
- Three targets on share of long term unemployment by age range (source INSEE 2011d)
- Three additional targets on part-time job proportion in employment (INSEE 2011d), vacancy rate (COE 2013) and the ratio of employed "looking for a new job" in the simulation (INSEE 2008)

4.2 Calibration Method

This fitness function is minimized at a horizon of 2,600 periods (each period corresponds to 1 week, then 2,600 periods correspond to 50 years in the reality). This choice of 50 years corresponds to a full career of the youngest agents and ensure that all agents have a fully simulated CV in the model and there is no bias related to the initialization phase of the model. To minimize our fitness function, we choose the evolutionary algorithm CMA-ES (Hansen and Ostermeier 2001), which is one of the most powerful algorithms to solve this kind of problem (Auger and Hansen 2012). Once the fitness function is minimized at the horizon of 2,600 periods in a steady state, we save all the states of the agents. This backup files will be the starting point for our analyzes of model variants.

4.3 Calibration Results

We obtain a median relative spread between the outputs of our model and the real targets of 16.5 %. These outputs are averaged over 240 simulations. All the targets and parameters values are given in our reference paper.[4]

5 Suppression of Fixed Duration Contract (FDC)

Once the model is calibrated, we aim to analyze the impact of a variant of labor policy in our model. In this paper, we test a Fixed Duration Contract (FDC) suppression. As in our model we only have two types of contract OEC and FDC (which is supposed to include all temporary contracts like interim and apprentice contract in our model), when we suppress FDC, only open-ended contracts (OEC) remain on the labor market.

We designed two simulation sets, one with OEC and FDC and one with OEC only. For each set, we start from the backup file mentioned above (cf. Sect. 4.2) and run 240 simulations (each one having a 4-year duration). The outputs presented in our results below are computed as averages of the last year over these 240 simulations.

[4] http://www-poleia.lip6.fr/~kant/WS1.6_MCR.pdf.

Table 1 Global impact by age of FDC suppression after 4 years

	With FDC	Without FDC	Gross spread
Global Unemployment rate	9.5 [9.426, 9.574]*	2.60 [2.52,2.67]*	−6.9**
15–24 Unemployment rate	20.4 [20.22,20.62]*	6.5 [6.30, 6.66]*	−13.9**
25–49 Unemployment rate	7.9 [7.82,7.96]*	1.6 [1.50, 1.61]*	−6.3**
50–65 Unemployment rate	8.6 [8.48, 8.65]*	3.2 [3.11, 3.29]*	−5.4**
Global Employment rate	62.6 [62.55,62.65]*	68.6 [68.52, 68.65]*	+6.0**
15–24 Employment rate	31.5 [31.36, 31.56]*	37.9 [37.73, 37.99]*	+6.4**
25–49 Employment rate	80.4 [80.36, 80.50]*	87.2 [87.13, 87.26]*	+6.8**
50–65 Employment rate	55.9 [55.83, 56.00]*	61.1 [60.32, 60.48]*	+5.2**

Employment rate: ratio of the employed over total population in the bracket 15–65
*Confidence interval at 95 %; **Significant at 1 % threshold with Student test

5.1 Global Impact by Age Group

As seen in Table 1, we observe a significant decrease of the unemployment rate for all age groups when we suppress the fixed duration contracts in the labor market. This decrease is particularly true for young people, as they are were more concerned by these short-term contracts.

In our model, the preference for the stability criteria (i.e. preference for OEC over FDC) is taken into account by the individuals in their utility evaluation and decision-making process [cf. Eq. (3)], but we observe the same significant decrease of the unemployment rate in our simulation even if we the labor supply does not depend on the expected duration of the contract.

5.2 Impact on Employee Turnover

To characterize churning, we measure the degree of mobility of our individuals with the employee turnover rates (given by the average of the entry and exit rates). The results are shown in Table 2.

With FDC the labor market is characterized by a high job turnover, especially among young people. When we suppress them, we observe a significant decrease of this job turnover, with less entry rate, but also less exit rate. This result suggests that the suppression of FDC reduces the global *churning effect* on the labor market. This reduction of mobility is much higher for the young (−59 % for the 15–24, −54 % for the 25–49 and −23 % for the seniors). When we suppress the FDC, the firm's *screening effect* obtained by short-term contracts remains, as more workers are fired at the end of the trial period. We also observe an increase of the number of layoffs for economic and personal reasons. These layoffs are used by firms to adjust their production in response to an idiosyncratic demand shock.

Table 2 Impact of FDC suppression on turnover rates by age range

	With FDC				Without FDC			
	Global	15–24	25–49	50–64	Global	15–24	25–49	50–64
Turnover rate	33.5	64.3	27.1	36.0	18.9	26.2	12.3	27.8
Entry in FDC rate	24.5	56.8	20.7	20.4	0.0	0.0	0.0	0.0
Entry in OEC rate	8.9	16.5	6.3	11.8	18.9	36.8	12.9	25.0
Exit rate	33.6	55.3	27.2	39.8	18.0	15.7	11.6	30.6
End of trial period rate	3.3	3.8	2.1	5.7	5.1	3.9	2.6	11.2
Layoff for eco. reasons rate	1.0	1.2	1.1	0.9	1.7	2.7	1.7	1.4
Layoff for other reasons rate	3.3	2.7	2.6	5.3	4.2	3.0	3.1	7.0

In the present version of the WorkSim Model the adjustment is not totally different between the case with FDC and the case without FDC. In the case with FDC, only the terminated FDC can be used for the adjustment, which therefore also relies on end of trial periods, personal and economic layoffs. In the case with only OEC, adjustment is also possible because layoffs of workers with a tenure less than 1 year do not undergo severance costs, but firms handle only salary costs during the advance notice period, and therefore firing is not on average very costly.

To summarize, the two cases offer the possibility to firms to adjust, and FDC do not provide a much higher buffer stock effect which would lower the unemployment in comparison to the only OEC case. The adjustments of firm production with end of trial period and layoff result in a global decrease of OEC duration (from 286 to 188 weeks). We observe a clear increase of the proportion of really short Open Ended Contracts: the proportion of contracts with an observed duration less than 1 month is multiplied by 2.4 and also by almost 1.4 for less than 6 months contracts. Hence, some precarity remains among the recently hired individuals, even though globally the exit rate is divided by 2.

5.3 *Impact on Flow Diagrams of Individuals*

To highlight this reduction of the *churning effect*, we present in this section the flow diagram[5] for all individuals with and without FDC.

[5] Each type of flow is measured in two ways. The percentage in brackets indicates the proportion of agents of a group who change state. This is actually a probability of transition to a state per period for a given agent. These probabilities are very low because they are calculated on a weekly basis but they show perfectly the relative probabilities to change state. The numbers associated with the arrows indicate the average number of agents who move from one state to another each period. These numbers of agents are given with the full-scale numbers (number of agents in the simulation multiplied by the scaling factor of 4700) and are stated in thousands of agents . The thickness of the arrow in the diagram shows the "flows strength" compared to the other flows.

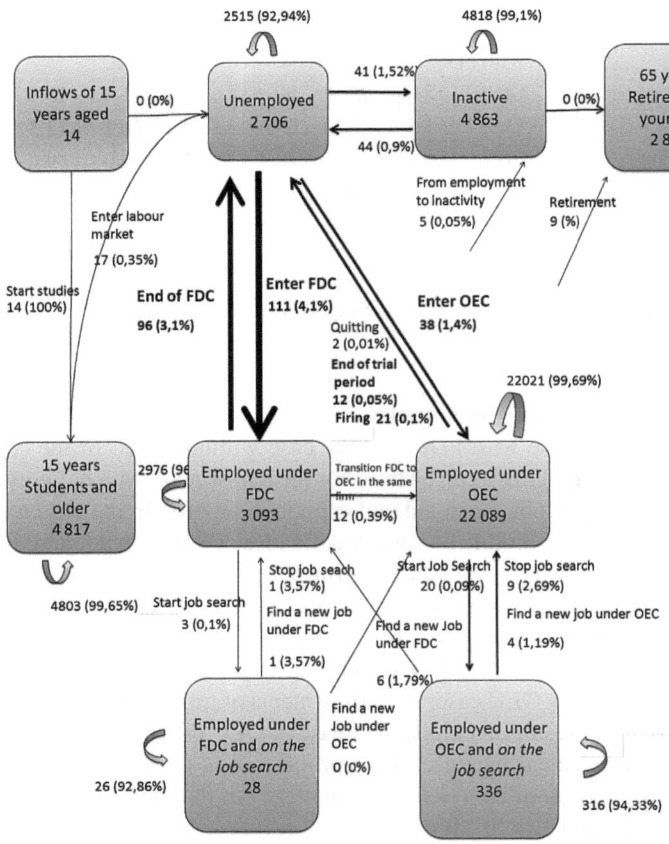

Fig. 2 Flow diagram of individual with FDC (in thousands)

In the diagram of all individuals with FDC (Fig. 2), the labor market is characterized by high rates of rotation between the status of "unemployed" and "Employee in FDC". Entry rates in FDC are more than three times greater than the rate of direct entry in OEC. Exit to unemployment is also a major stream, the second in size. The conversions of FDC into OEC represent only 11.8 % of the exits, the others persons go to unemployment. Therefore, an important part of agents alternate short periods of temporary work with periods of unemployment. Besides these "precarious" status, the majority of employees are employed in very stable OEC, even if some of these permanent employees are always looking for another more attractive job in OEC (transition to the "on the job search" state). When we suppress FDC (Fig. 3), we observe an increase of the flow strength between Employment and Open-ended Contract. The important part of agents who alternated short periods of temporary

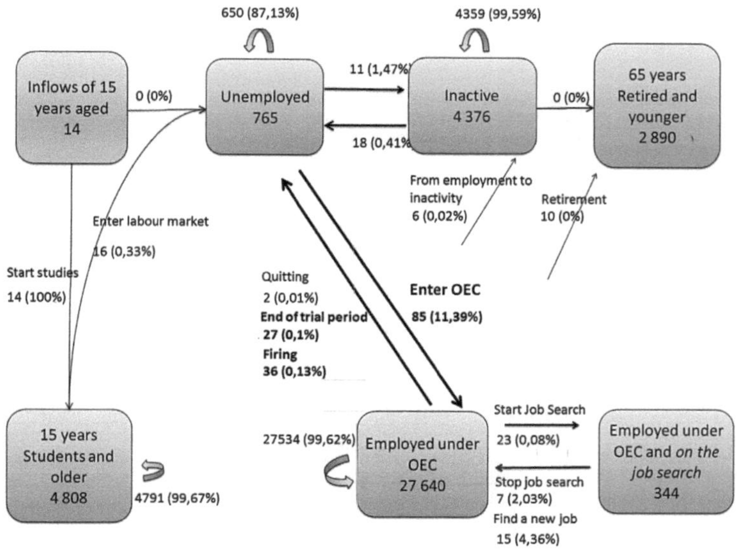

Fig. 3 Flow diagram of individual without FDC (in thousands)

work with periods of unemployment when FDC was available is partly replaced by agents who alternate unemployment and very short employment episodes in OEC.

5.4 Impact on the Beveridge Curve

This decrease of employee turnover results in less frictional unemployment. Indeed, with the suppression of very short term contracts the job turnover decreases and the mean duration of employment spell increases. Hence, we observe a net decrease of the number of firms searching for employees and of the number of unemployed searching for jobs per period. This explains the net shift of the Beveridge[6] curve towards the origin when we suppress FDC, as shown in Fig. 4.

This observation is consistent with the results obtained by Bentolila et al. (2010) in their comparison of the labor market of France versus Spain. They show the large outward shift in the Beveridge curve of the Spanish labor market during the crisis in 2010, because this labor market was characterized by a strong duality between OEC and FDC and by a high share of individuals employed in FDC. This outward shift of both unemployment rate and vacancy rate is an indicator of frictional unemployment due to the *churning effect*.

[6]The Beveridge curve is obtained by shifting the global demand of our model from 50 % to 200 % of the base value.

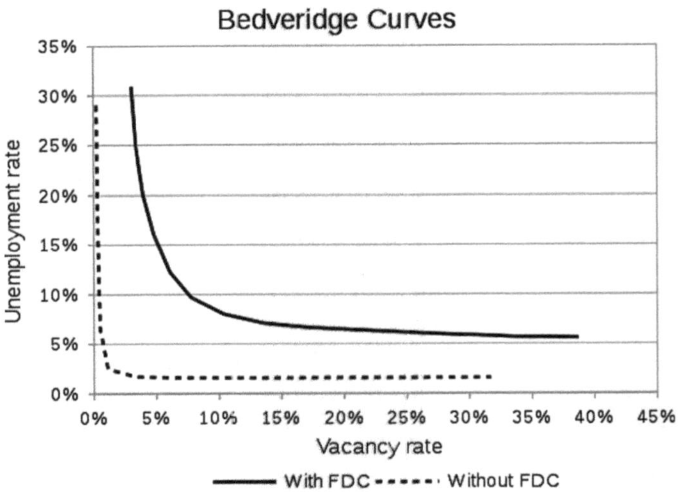

Fig. 4 Beveridge curves with and without FDC

5.5 *Sensitivity of FDC Suppression to the Volatility of Firms Demand*

In our model, the dynamics of the firms demand has an impact on their decision process concerning jobs creation and destruction (cf. Sect. 2.1). We aim to test in this section the sensitivity of our results to the volatility of firm demand. We run 48 simulations with FDC and 48 simulations without FDC with values of volatility of firms market share σ_{MS}[7] each period between 0 and 0.07 (i.e. between 0 % and 200 % of the calibration value 0.035).

The more the volatility of firms demand increases, the more firms are likely to hire and fire employees each period in order to adjust their production level to their demand level. Figure 5a on employee turnover shows two differences between the two cases. First employee turnover is always lower without FDC as we have seen in Table 2. Second the rate of increase in turnover rate when volatility rises is higher with FDC, due to the higher flexibility of short term contracts. Figure 5b shows that when the volatility is high, the unemployment rate for young people is high with FDC and low without FDC. This high volatility is the case of the base simulation (Sect. 5). The high volatility associated with a high turnover then causes an important *churning effect*. When the volatility rate is very low, the hierarchy of unemployment rates for the young is reversed. The turnover rate is very low and

[7]As in Eq. (1) in Sect. 2.1

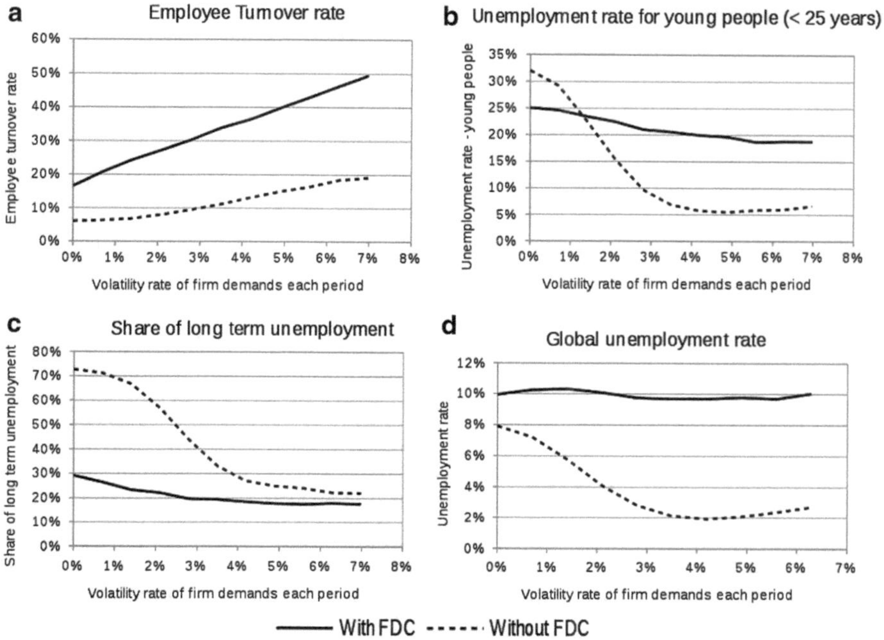

Fig. 5 Sensitivity to volatility of firm demands with and without FDC

it becomes very hard for the young people and more generally for the individual with little human capital to enter into employment, and this increases long term unemployment (Fig. 5c). This difficulty is smaller when FDC are available since inexperienced workers and individual unemployed have a higher probability to get a job, mainly an FDC, which increases their experience and raises the probability to access an OEC later.

This highlights the existence of a positive *stepping stone* effect when FDC are allowed and volatility is low, which does not occur when only OEC exist. The degree of volatility of demand determines two opposed scenarios of the effects of FDC suppression. When volatility is low, the *stepping stone effect* overcomes the *churning effect*.[8] When the volatility is high the stepping stone effect disappears and there remains only a churning effect, which moreover is strengthened. At the level of the global unemployment rate, the reversal of hierarchy does not appear since the stepping stone is not so important for the older workers (Fig. 5d).

[8]This is the case studied by Ballot (1981, 2002), in which the firms represented sectors, which have a lower demand volatility than individual real firms.

6 Discussion

To sum up the WorkSim Model enables us to study different mechanisms that are at the core of the labor market policy experiment and produce opposed effects on unemployment, mainly the *churning* and the *stepping stone effect*. In the base simulation, which corresponds to the calibrated values for the French labor market in 2011, with a relatively high volatility of idiosyncratic firm demand and employee turnover, the *churning effect* overcomes the *stepping stone* effect and the net effect of the suppression of FDC is an important decrease in unemployment. If the volatility of demand is very low, the turnover is also low, vacant jobs are very few, and the suppression of FDC makes the integration of young people and unemployed very difficult. The global unemployment rate remains lower but the unemployment rate for young people is higher.

The net effect of the FDC suppression may also be influenced by the *buffer stock effect* that we do not endogenize completely, because the firm decision to decide between an OEC and FDC is partly exogenous: first an OEC or FDC is drawn from a exogenous probability (calibrated) but then the firm decides to actually create the job following an endogenous decision process. However this buffer stock effect might not be very important, as some theoretical work suggest (Bentolila et al. 2010), and also because the French legal framework modeled in WorkSim allows to terminate an OEC during the trial period and to fire employees with less than 1 year without severance payment.

An extension of the study would consider some other firing costs such as the bargaining of supra legal severance payments, the necessity of obtaining authorization for collective economic layoffs and finally the reputation cost that can be incurred when laying off. Moreover, we have assumed that temporary jobs are integrated in the FDC, but they would deserve to be modeled with their specificity. Finally, we will improve the choice mechanism between OEC and FDC, so the firm will perform it in a endogenous way.

References

Auger A, Hansen N (2012) Addressing numerical black-box optimization: CMAE-ES. In: LION 6, Paris, France, 16–20 January 2012

Ballot G (1981) Marché du travail et dynamique de la répartition des revenus salariaux. Thèse pour le doctorat d'Etat d'Economie, Université Paris X-Nanterre

Ballot G (2002) Modeling the labor market as an evolving institution: model artemis. J Econ Behav Organ 49(1):51–77

Bentolila S, Bertola G (1990) Firing costs and labour demand: how bad is eurosclerosis? Rev Econ Stud 57(3):381–402

Bentolila S, Cahuc P, Dolado JJ, Le Barbanchon T (2010) Two-tier labor markets in the great recession: France vs. spain. Technical report, Discussion paper series//Forschungsinstitut zur Zukunft der Arbeit

Berche K, Hagneré C, Vong M (2011) Les déclarations d'embauche entre 2000 et 2010: une évolution marquée par la progression des CDD de moins d'un mois. Acoss Stat (143)
Berson C, Ferrari N (2013) Réduire la segmentation du marché du travail par des incitations financières? DG Trésor 2013/04, DG Trésor
Blanchard O, Landier A (2002) The perverse effects of partial labour market reform: fixed-term contracts in france. Econ J 112(480):F214–F244
Booth AL, Francesconi M, Frank J (2002) Temporary jobs: stepping stones or dead ends? Econ J 112(480):F189–F213
Bucher A (2010) Hiring practices, employment protection and temporary jobs. Paper 2010–13, TEPP Working Papers
COE (2013) Emplois durablement vacants et difficultés de recrutement. Rapport du Conseil d'orientation pour l'emploi, p. 28
DARES (2012) Les mouvements de main-d'oeuvre en 2011. DARES Analyses (71)
De Froment C (2012) Pour une flexibilité responsable. Sociétal 75:112–119
Faccini R (2008) Reassessing labor market reforms: temporary contracts as a screening device. Working paper, European University Institute
Hansen N, Ostermeier A (2001) Completely derandomized self-adaptation in evolution strategies. Evol Comput 9(2):159–195
INSEE (2008) Population active à la recherche d'un autre emploi (PARAE)
INSEE (2011a) Entreprises selon le nombre de salariés et l'activité en 2011
INSEE (2011b) Taux d'activité selon le sexe et la configuration familiale en 2011
INSEE (2011c) Taux de chômage par âge en 2011
INSEE (2011d) Une photographie du marché du travail en 2011
INSEE (2013a) Fiches thématiques - Synthèse des actifs occupés - Emploi et salaires - Insee Références - Édition 2013
INSEE (2013b) L'emploi dans la fonction publique en 2011
Jauneau Y, Nouel de Buzonniere C (2011) Transitions annuelles au sens du BIT sur le marché du travail. INSEE
Mortensen D, Pissarides C (1994) Job creation and job destruction in the theory of unemployment. Rev Econ Stud 61(3):397–415
Omicini A, Ricci A, Viroli M (2008) Artifacts in the A&A meta-model for multi-agent systems. Auton Agents Multi Agent Syst 17(3):432–456
Phelps E (1970) Microfoundations of employment and inflation theory. Macmillan, London
Boeri T (2011) Institutional reforms and dualism in european labor markets. Handb Labor Econ 4:1173–1236

Shadow Economy and Wealth Distribution

Nuno Trindade Magessi and Luis Antunes

1 Introduction

Shadow economy effectively misleads the government in national accounts by decreasing tax revenue, with consequences on government's ability to provide public services and hence increasing the nation's debt. Shadow economy includes economic activities and respective incomes that are not under the government regulation and taxation. "Noncompliance shifts real resources from honest taxpayers to dishonest evaders and tax liabilities from present to future generations" (Feige and Cebula 2011). In this sense, there is a shift from legal and regulated economy to the underground economy.

In 2008, estimations were done and found underground activity to total as much as $2 trillion in USA (Feige 2011). This huge number reveals and measures the importance of the phenomenon. However, from the inverse side, shadow economy can generate value to formal economy since the value accumulated by non-compliers economic agents is spent or invested in the official economy (Nikopur et al. 2008). Clearly, there is an allocation of resources between both economies.

Being illegal and non-complying, unofficial economy often represents an opportunity for poor individuals, who do not comply to have access a cheaper resources and doing their businesses that leads to an increase in their incomes. The consequence to society is to have more fairness in the wealth distribution (Nikopur et al. 2008).

The main goal of this article is to contribute to the discussion about the relation of both economies, but focused on shadow economy. For instance, we purpose a simulation model to analyse the fairness of the wealth distribution where shadow

N.T. Magessi (✉) • L. Antunes
GUESS, LabMag, Universidade de Lisboa, Lisboa, Portugal
e-mail: nmaggessi@hotmail.com; xarax@ciencias.ulisboa.pt

F. Amblard et al. (eds.), *Advances in Artificial Economics*, Lecture Notes in Economics and Mathematical Systems 676, DOI 10.1007/978-3-319-09578-3_14

economy shift resources to formal economy. In this simulation we can see Pareto's Law (Pareto and Page 1971) working on, by taking into account the shifts from shadow to regulated economy.

This article is organised as follows: in the next section, we review the literature about shadow economy. In Sect. 3 we summarize the state of the art of agent-based-modelling of the tax evasion problem. Section 4 describes the model and its specific subject-related mechanisms. Section 5 discusses the results we obtained by running simulations with this model and confronts them with other studies. Finally, on Sect. 6, we draw our conclusions and introduce future steps for this research.

2 Reviewing Shadow Economy literature

As it was described earlier, shadow economy happens because economic agents want to be free on their activity and avoiding government meddling. This happens because such agents want to establish pure market rules, since government interference brings always associated constraints precluding maximizing their profits. The principal and biggest constraint is obligatory payments to government entities. These compulsory payments, like social contributions and taxes, typical results on the practice of their evasion by economic agents. Tax evasion is a very rich field in literature. For a deep, broad, and structured coverage of the field, see Cowell (1990), Andreoni et al. (1998), Alm (1999), Franzoni (2000) and Slemrod (2007). The literature's theoretical models had the intent of identifying variables that explain tax evasion behaviour. Unfortunately, the assumptions used in modelling behaviours, and the specifications of the models, often lead to contradictory outputs. These "theoretical ambiguities" unleash further empirical analysis to examine the determinants of tax evasion (Feige and Cebula 2011). Besides behaviour, measuring tax evasion is another issue that took attention from literature. Studies presented recently (Feige and Cebula 2011; Feige 2011) prove the actuality and pertinence of this subject. Tax evasion has becoming highly sophisticated, using innovative schemes and processes, taking into account the recent developments done by tax authorities in processing and crossing data.

In 2007, Chen (2007) identified the existence of three approaches in the relation between official and shadow economies: dualism, structuralism and legalism. Dualism establishes that shadow economy has few connections to official economy and operates separately. Its hypothesis sustain that regulation had segmented the market, as a derivation from the rigidities of official economy (Chen 2007). These few connections happen since both economies share some common factors, like unemployment, corruption level or monetary mass, allowing the transference of resources between these economies (Chen 2007).

On the other hand, structuralism assumes that official and shadow economies are intrinsically connected. This means that some agents from the official economy encourage relations with shadow economy in order to decrease input costs. Agents meet their interests and consequently the shadow economy is used to expand the official economy.

Instead, legalism establishes a relation between the shadow economy and the regulatory environment of the official economy, outside the scope of the agents' actions (Chen 2007). There is a collusion of interests between economic agents and government in the regulated official economy.

In 2008, Nikopur et al. (2008) presented another perspective. They suggest that to understand the consequences of shadow economy we should focus on the nature of the relation with formal economy. For these authors, what is important is to know whether, in the relationships between both economies, substitute effects such the passage of productive activities overcome complementary ones, like economic growth. When both economies complement instead of competing each other, the shadow economy stimulates the growth of official economy. The authors justified this claim with the value added in the shadow economy, which is subsequently transferred to the official economy. However, if the competition between both economies prevails, unfair competition affects negatively the allocation of resources. So Nikopur et al. (2008) has shown us, that there are positive and negative impacts of shadow economy in official economy.

3 Reviewing Multi-Agent Literature About Tax Evasion

Tax evasion studies have proliferated over the last few years, based on the behaviour of agents involved on it and measuring its impacts on official economy. Those studies have their core on neoclassical economic theory. The developed models attempted to explain behaviour biases in terms of rational choice. Many of those studies have their focus on measurements and are based on the morality of economic agents, defined as "intrinsic motivation" or "internalized willingness" to pay taxes (Miguel et al. 2012).

Recently, a small group of studies used agent-based models to explain tax evasion, through simulation, at aggregate level. Those studies expose tax evasion not only restricted to the degree of deterrence and incorporating other effects and dimensions, like social interaction. One of the interesting works and more related with shadow economy is one of EC* series (Antunes et al. 2006, 2007). These models were done progressively by adding complexity over the standard economic model. The most notable development of this sequence of models was the introduction of independent tax inspectors with autonomous decision-making. The authors of Antunes et al. (2007) also suggested that unpaid indirect taxes must arise from a collusion of interests between buyers and sellers implicitly done in shadow economy. The latter study directly focuses on one of the main necessary conditions for the growth of shadow economy: the collusion of agent's interests.

More recently, Bloomquist (2011) analyzed tax compliance for small business and interpreted as an evolutionary coordination game. He calibrated the model with real data from behavioural experiments. Additionally, Miguel et al. (2012) used a different algorithm based on four different decisional mechanisms: expected utility maximisation, social network structure, decisional heuristics and heterogeneity of tax motivations and morale.

Agent-based social simulation generates positive expectations about the future of this field of knowledge. The biggest identified problem with other methodologies is the incapacity to integrate the relevant dimensions involved.

4 Wealth Distribution Model for Shadow Economy

As it was described above, shadow economy is normally alleged as an instrument to bring more equity to the wealth distribution in society, since economic agents have access to resources that otherwise would be withheld from them (Nikopur et al. 2008). We propose to simulate this situation in order to understand if this really happens.

4.1 SEWD Model Parameters

The Shadow Economy Wealth Distribution model has a group of parameters that influence wealth distribution and are controlled by the researcher. The parameter designated by percent best country (ρ) determines the initial density of areas (countries) where the shadow economy is seeded with maximum installed capacity (ω). The percent best country has implications on defining the quantity of countries that are permissive to informal economy. The parameter ω designated as maximum installed capacity represents the maximum value of resources available in that country.

The shadow economy growth cycle (ϕ) determines how often gross product grows in each country where shadow economy prevails. This parameter represents the inverse side of formal economic cycle. If this cycle is long it means that non-complier agents find it more profitable to do their businesses on shadow economy. This happens because agents are not punished in evading taxes, for example when a country has a high fiscal charge. Another parameter is the growth of gross domestic production (g) and it determines how much money is incremented in shadow economy at each economic cycle.

The parameter non-complier agents (η) determines the initial number of non-complier agents who are susceptible to do their activities inside of shadow economy. The minimum expected activity (m) represents the shortest number of ticks that a non-complier agent can possibly do his/her own business in shadow economy. On the other hand, the maximum expected activity (M) is the longest number of ticks that a non-complier agent can possibly perform his or her activity under the premises of the shadow economy. The maximum money laundering (ψ) parameter sets the highest possible amount of income that an agent could invest or spend per tick. The maximum access parameter (π) is the furthest possible country that any non-complier agent could have access. The justification for this parameter is the fact that sometimes agents face restrictions in having access to informal economy in other countries, like the specific authorizations to be residents.

4.2 SEWD Model Description

This model is designed to reveal a worldwide artificial society where shadow economy and formal economy cohabit together. The model is composed by countries where the shadow economies can proliferate and other countries in which the official economy is predominant, and where the development of shadow economy is conditioned. In our simulator, patches represent countries where exist economic activity. A dark patch represents a place where shadow economy could not proliferate, since it is controlled by government rules and where official economy prevails. On the contrary, a light yellow patch symbolises a country where it is possible to develop clandestine activities without restrictions. The variation of the colour depends of the predominance of shadow economy in each country. The patch changes dynamically from yellow to dark and vice versa revealing the fight actions done by government authorities against shadow economy and money laundering activities. Each country, where informal activities are available has an amount of resources available that allows a specific production (p). A production that has a certain capacity depending of its factors. This capacity is settled randomly and limited superiorly by ω. In this sense, production comes in function of the percent best country and the maximum installed capacity (1).

$$p = f(\rho, \omega) \tag{1}$$

$$p_t = p_{t-1}(1 + g)^t \tag{2}$$

Additionally, the amount of resources and consequent production can grow on each country which translates into the possibility of shadow economy growth (2). Non-compliers agents collect the income generated by their production using the available resources from the countries where shadow economy is available and spend or invest it on official economy, in order to increase legally, their wealth. This process is called money laundering and it is established by randomly a value (δ) less than a settled maximum for this procedure ($\delta < \psi$). The income (γ) comes from production, where each agent accumulates in each tick generating the agent wealth. We assume that the income of each agent is equal to the gross domestic production per capita in shadow economy.

The model begins with a randomly-settled, roughly equal wealth distribution. Non-complier agents are divided according their initial wealth, in three classes: rich, medium class and poor. Then an agent wanders around a country where shadow economy is available gathering as much income as he cans (3).

$$\gamma_t = \max\ (p_t, \pi_t) \tag{3}$$

where

$$\pi_t = f\ (\pi_{t-1}) \tag{4}$$

They attempt to move in the direction where most of the unrestricted resources lay and where they allowed to go. This permission is given by the parameter that rules the maximum countries (patches) that an agent could have access (4). This means that agents could be restricted on their movements to a specific country economy. In each time tick, each agent spends or invests a certain income on the official economy that he or she was accumulating through shadow economy. Consequently, every run the wealth (w) is calculated as follows (5):

$$w_t = w_{t-1} + \gamma_t - \delta_t \tag{5}$$

Agents also have a shadow activity expectancy, according to the wealth they want to achieve. The expected activity (ea) is defined as (6):

$$ea = m + (M - m) \tag{6}$$

where the difference (M − m) is settled randomly by software.

When agents activity runs out ($t >= M$), where t is time or when they do not launder money because $w_t < 0$, they simple get out of activity without moving their business to another agent. However, a new non complier agent starts with a random quantity of income ranging uniformly from the poorest to the richest agent, in activity. This means that there is no inheritance of wealth.

To analyse the fairness of the wealth distribution, we have drawn the Lorenz curve, a tool normally used in these circumstances. We have ranked non-complier agents by their wealth and then we have plotted the percentage of them that owns each percentage of the wealth. We ranked the agents in order of their wealth, from the greatest to the least: the poorest agent would have the lowest ranking of 1 and so forth. Then we have plotted the proportion of the rank of an agent on the y-axis and the portion of wealth owned by this particular agent and all the agents with lower rankings on the x-axis. For example, agent alpha with a ranking of 20 (20th poorest in society) would have a percentage ranking of 20 % in a society of 100 agents. The corresponding plot on the x-axis is the proportion of the wealth that this agent with ranking 20 owns along with the wealth owned by the all agents with lower rankings, from 1 to 19.

A straight line with a 45° angle at the origin (or slope of 1) is a Lorenz curve that represents perfect equity, meaning that everyone holds an equal share of the available wealth. On the other hand, should only one individual hold all of the wealth in the population (i.e. perfect inequity), and then the Lorenz curve will be a backwards "L" where 100 % of the wealth is owned by the least possible percentage proportion of the population.

For a numerical measurement of the fairness of the distribution of wealth, the Gini coefficient is derived from the Lorenz curve. To calculate the Gini coefficient, we first have found the area between the 45-degree line of perfect equality and the Lorenz curve. Secondly, we divided this quantity by the total area under the 45-degree line of perfect equity, which is always 0.5. If the Lorenz curve is the 45-degree line then the Gini coefficient would be 0, meaning that there is no area between the Lorenz curve and the 45° line. If, however, the Lorenz curve is a backwards "L", then the Gini coefficient would be 1. Hence, equity in the distribution of wealth is measured on a scale of 0–1.

5 Relevant Results

The results reported in this section were obtained conducting the described experiments using version 5.0.4 of the NetLogo framework (Wilensky 2012). NetLogo is a programmable modelling environment for simulating natural and social phenomena. It is particularly well suited for modelling complex systems developing over time.

At this stage of research, we are still doing more experiences and getting insights on how the several parameters impact the system dynamics. So, in this section we will only hint, the obtained results so far, and present a brief analysis. Our world has 500 of one thousand economic agents ($\eta = 500$).

First of all we questioned if the growth of production through shadow economy affects negatively the distribution of wealth in global society. The obtained results show us that the growth of production in shadow economy is a source of inequity for society, but attenuated if the cycle of shadow economy becomes larger (see Figs. 1, 2 and 3). As we can see, the number of poor agents rises significantly and this class becomes predominantly worldwide. The Lorenz curve moves away significantly and the Gini Index increases considerably compared to Fig. 1. However, if the economic cycle of shadow economy has a greater duration than it had in Fig. 1 the inequity worldwide decreases. Theoretically, a large or prolonged cycle of economic growth in shadow economy comes from a decrease of growth in official economy. A decrease motivated by the increase of taxes and social contributions in order to the governments control their budgets and respective national accounts. Or to

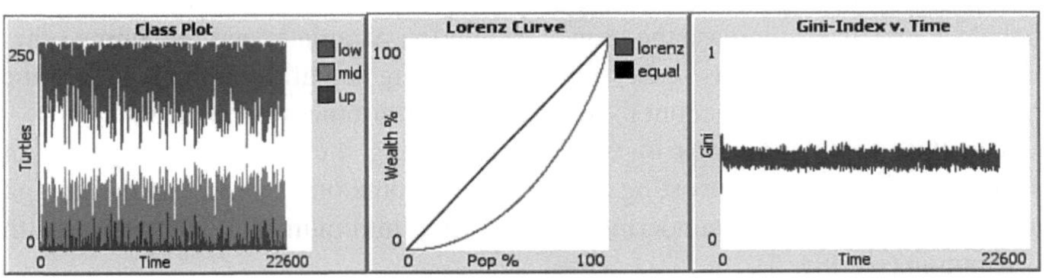

Fig. 1 Official economic growth scenario—Gini-Index [38–47 %]

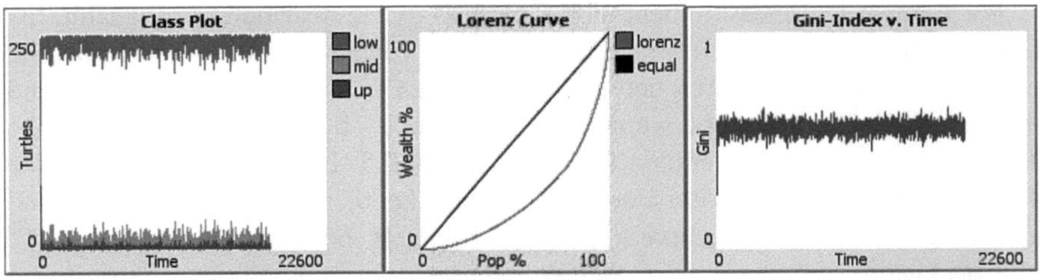

Fig. 2 Increase on the growth of shadow economy production—Gini-Index [42–61 %]

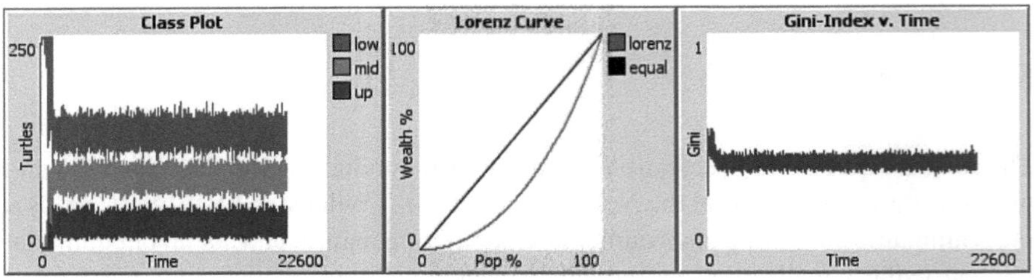

Fig. 3 Increase of production growth and on the shadow economic cycle—Gini-Index [35–43 %]

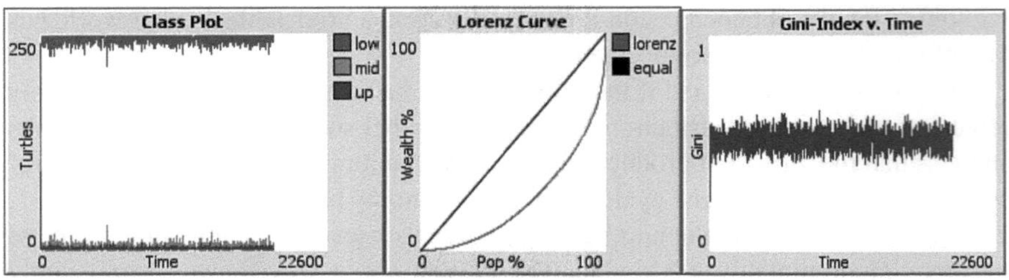

Fig. 4 Increase of money laundering parameter—Gini Index [42–69 %]

implement new rules, full of requirements and constraints that destroys the market's value.

The second simulation was to check what happens if agents increase the investment in official economy, or in other words if they increase their money laundering activities. Results demonstrate a considerable increase on the inequity inside of global society, if we increase the money laundering parameter and remaining other parameters constant. In this sense, money laundering activities put on evidence the Pareto's law taking into account these specific circumstances (see Fig. 4).

The third simulation was to verify if increasing the production capacity of shadow economies by reinvesting the obtained income on it, sustained by its high growth and where money laundering is settled on its minimum allows a better wealth distribution (see Fig. 5).

According this simulation we can see that the reinvestment in shadow economy decreases slightly the wealth distribution asymmetries worldwide. Both, Lorenz

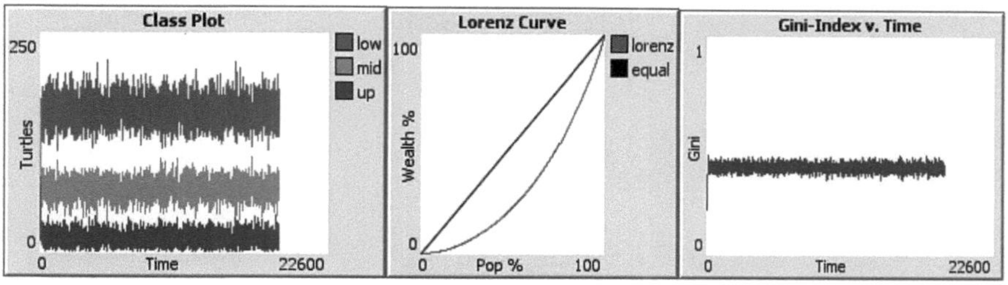

Fig. 5 Double investment in the capacity of production on shadow economy—Gini Index [33–43 %]

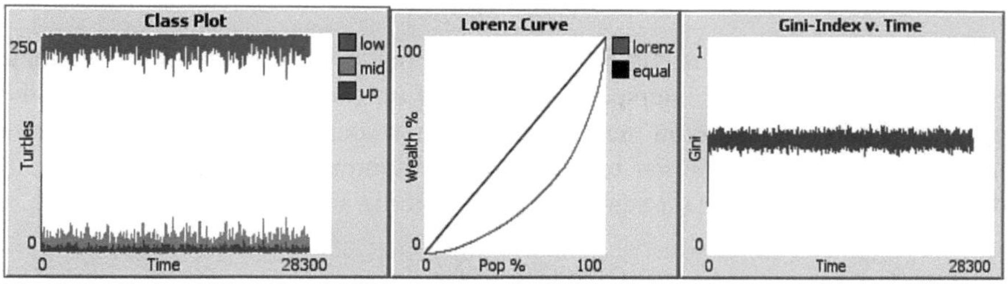

Fig. 6 Shadow economy as a closed economy—Gini-Index [46–54 %]

curve and Gini Index reveal this conclusion comparative to the scenario of Fig. 1. We increased the economic cycle of shadow economy and there is no substantial difference. Comparing with the scenario of Fig. 2, the reinvestment on shadow economy capacity avoids the initial deterioration on wealth distribution worldwide. This simulation suggests that investing in shadow economy capacity contributes more than high economic cycle for fairness at worldwide.

Another important result is to verify the case where shadow economy tends to be a closed economy. In other words, the case where agents face restrictions on their movements worldwide and consequently agents cannot take advantage from an open economy. In this sense, the competition is restricted to each country and becomes more intensively because of resources scarcity (see Fig. 6).

Simulation reveals that reinvesting in production on a closed economy intensifies competition among agents in a restrict country and impacts negatively in the distribution of wealth. The action has similar consequences if money laundering had been increased.

Finally, the question, is it possible for shadow economy contribute to more fairness worldwide? The results reveal that this is possible under some conditions. The conditions are: (1) agents can easily fulfill their wealth expectations in short periods of time using shadow economy, which is when $M - m$ is lower; (2) shadow economy is completely disseminate by all countries; (3) agents face no restrictions in moving ahead; and (4) shadow economy face a long economic cycle of growth (see Fig. 7).

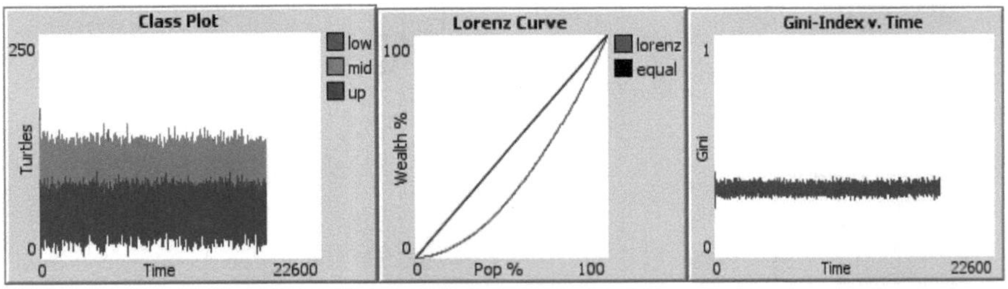

Fig. 7 Increase in percent best country and maximum access activity together with a short expected activity in shadow economy—Gini Index [20–32 %]

These results show us that if agents find out good conditions to do their businesses in a short and interspersed periods of time and during a long economic cycle of shadow economy in many countries, then society becomes more fairness in what concerns distribution of wealth. The asymmetry in wealth distribution decreases and middle class assumes preponderance in society.

Conclusion and Future Work

Shadow economy has been studied fundamentally in what concerns the value it represents and the matters behind of it. However shadow economy is never looked as economy like the official one. In this article, we analyzed the effect of shadow economy in the distribution of wealth worldwide. Preliminary results show us that fairness is a concept that can be achieved if the shadow economy is completely disseminated worldwide where agents have full access to its resources and fulfil their wealth expectations in a short period of time. Money laundering, independently of stimulating the growth of official economies is a mechanism that provokes the spread of inequity worldwide inducing evidence of Pareto's law. In this sense, shadow economy could contribute for official economy growth, however the beneficiaries is a small fringe of society, the wealthy.

Future work intends to include the penalties that agents suffer on their wealth if they are discovered.

References

Alm J (1999) Tax compliance and administration. In: Bartley Hildreth W, Richardson JA (eds) Handbook on taxation. Marcel Dekker, New York, pp 741–68

Andreoni J, Erard B, Feinstein J (1998) Tax compliance. J Econ Lit 36(2):818–60

Antunes L, Balsa J, Moniz L, Urbano P, Palma CR (2006) Tax compliance in a simulated heterogeneous multi-agent society. In: Sichman JS, Antunes L (eds) MABS 2005. LNCS (LNAI), vol 3891. Springer, Heidelberg

Antunes L, Balsa J, Coelho H (2007) Agents that collude to evade taxes. In: Proceedings of the 6th international joint conference on Autonomous agents and multi-agent systems (AAMAS 2007)

Bloomquist K (2011) Tax compliance as an evolutionary coordination game: an agent-based approach. Public Financ Rev 39:25

Chen M (2007) Rethinking the informal economy: linkages with the formal economy and the formal regulatory environment. Working paper n. 46. United Nations, Department of Economic and Social Affairs

Cowell FA (1990) Cheating the government: the economics of evasion. M.I.T. Press, Cambridge

Feige EL (2011) New estimates of U.S. currency abroad, the domestic money supply and the unreported Economy, MPRA Paper 34778, University Library of Munich, Germany

Feige EL, Cebula R (2011) America's unreported economy: measuring the size, growth and determinants of income tax evasion in the U.S. Forthcoming in: crime, law and social change no. April 2012

Franzoni LA (2000) Tax evasion and tax compliance. In: Bouckaeri B, De Geest G (eds) Encyclopedia of law and economics, vol 4. Edward Elgar, Cheltenham, pp 52–94

Miguel F, Noguera J, Llàcer T, Tapia E (2012) Exploring tax compliance: an agent based simulation. In: Troitzsch KG, Möhring M, Lotzmann U (eds) 26th European Conference on Modelling and Simulation (ECMS)

Nikopur H, Habibullah M, Schneider F (2008) The shadow economy Kuznet's curve panel data analysis, MPRA Paper No. 12956. Posted 29 Jan 2009

Pareto V, Page AN (1971) Translation of Manuale di economia politica (Manual of political economy). A.M. Kelley, ISBN 978-0-678-00881-2

Slemrod J (2007) Cheating ourselves: the economics of tax evasion. J Econ Perspect 21(1):25–48. doi:10.1257/jep.21.1.25.Abstract

Wilensky U (2012) NetLogo. Center for Connected Learning and Computer-Based Modeling, Northwestern University, Evanston

Distribution Effects of Extortion Racket Systems

Klaus G. Troitzsch

1 Introduction

Extortion racket systems like the ones known from mafia-like organisations in Southern Italy ('Ndrangheta, Cosa Nostra, Camorra) but also elsewhere in the world participate in the economic process in a peculiar way as they sell protection against their own offences, but also against the offences of others, thus paralleling the state which is the firstborn agency in charge of public security. This business, belonging to the shadow economy, does not contribute to the official GDP but is estimated to have a size of 16 % of the official GDP (Pinotti 2012, p. 18) (see also Spina 2008).

The model presented here takes a system perspective, leaving individual decision making processes for further publications. In a way it extends (Troitzsch 2010, pp. 60–62) which used the dynamics of legality and illegality as an example of communication and interpretation processes among agents in artificial societies, but it concentrates on the results of these interactions, leaving the details of decision making processes open for the moment and postponing them for future papers of the GLODERS project. Thus the research questions of this paper are the following:

- How does an extortion racket system like the mafia affect the distribution of wealth of enterprises and criminals?
- How does the attitude of the public towards the mafia affect the propensity of enterprises to contribute to the fight against the mafia?
- Which of the behaviour traits of criminals, enterprises, the public and the state affects the society as a whole most?

K.G. Troitzsch (✉)
Institut für Wirtschafts- und Verwaltungsinformatik, Universität Koblenz-Landau, Koblenz-Landau, Germany
e-mail: kgt@uni-koblenz.de

F. Amblard et al. (eds.), *Advances in Artificial Economics*, Lecture Notes in Economics and Mathematical Systems 676, DOI 10.1007/978-3-319-09578-3_15

The model represents a fictitious region whose economy simply consists of a number of firms (of any kind), a number of people consuming the goods and services (of any kind) of these firms, of a police force, and finally of persons who engage themselves in extortion rackets, asking the firms for extortion money ("pizzo") and eventually punishing them when they refuse to pay. The "addio-pizzo" movement (Spina 2009) is not explicitly modelled as an agent but has considerable influence on what happens in the model.

2 The Model

The four main agent types of the model are the following:

Shops serve for providing the population with goods and services of any type. To keep things simple, these goods and services are not detailed. Shops receive payment from their customers and bear the fixed and variable costs of their business by paying into a funds which is evenly distributed to the population which is considered to provide the shops with the necessary supply (both in goods and services)—to keep things simple here, too, the households of the population receive an equal share of the overall period income of the shops, as if all of them served as suppliers and workers for the shops equally.
Shops are often approached by criminals for "pizzo" which they can refuse at the risk of being severely punished. They can avoid this with a certain propensity to call the police (`denunciation-propensity`), denouncing the extorters and having them prosecuted. When a shop decides to denounce an extorter it joins an "addio-pizzo" movement and makes this fact known to everybody.
In case they are approached by more than one extorter at the same time—this happens mainly in the initial phase and represents what might have happened in the real world when mafia-like organisations first came into being—the shops decide whom to pay to, and the successful extorter will then protect the shop against the rivalling extorters. When due to extortion and punishment the asset of a shop falls below zero, it is closed and does not participate in the trading process until it is compensated from a funds filled by the confiscated wealth of the extorters.

Consumers choose a shop for purchasing their goods. They have a certain propensity (`addio-pizzo-threshold`) to choose shops belonging to the "addio-pizzo" movement. If their current shop is closed, they choose another, preferably belonging to the "addio-pizzo" movement and preferably having only a small number of customers (the reason for this is twofold: they want to avoid crowded shops, and shops with only few customers should get a chance to prosper).

Extorters start their career as individual criminals and approach the nearest reachable shop, asking for a "pizzo" which is a certain proportion (either `extortion-level-low` or `extortion-level-high`, depending on the type the extorter belongs to) of its revenue per period. When the shop refuses and the police fails to successfully prosecute the extorter, the latter punishes the shop, taking away a certain proportion of all its assets (either `punishment-severity-low` or `punishment-severity-high`, again depending on the type the extorter belongs to—i.e. there can be up to four types of extorters, low-low, low-high, high-low and high-high). It is understood that all assets of a shop are easily convertible into money, there are no physical assets which could be destroyed. If the police hinders the extorter from punishing, the extorter is brought to jail for a certain number of periods, and all its assets are confiscated and transferred into a funds from which in turn punished shops can be compensated[1] (following a first come–first served principle). If several extorters approach the same shop at the same time, one of them is selected by the shop to protect the shop against rivalling extorters, and the latter subordinate to the former, forming a family and eventually a hierarchy, for instance in case the successful extorter is already subordinate to someone else; if any rivalling extorters already belong to families, the family hierarchy is not changed.

Police try to prosecute a denounced extorter and are successful with a certain probability (`prosecution-propensity`)—which is their only role within the model.

Figure 1 shows the interface of the NetLogo (Wilensky 1999) realisation of the model.[2] For mainly illustrative purposes the region is subdivided into a number of

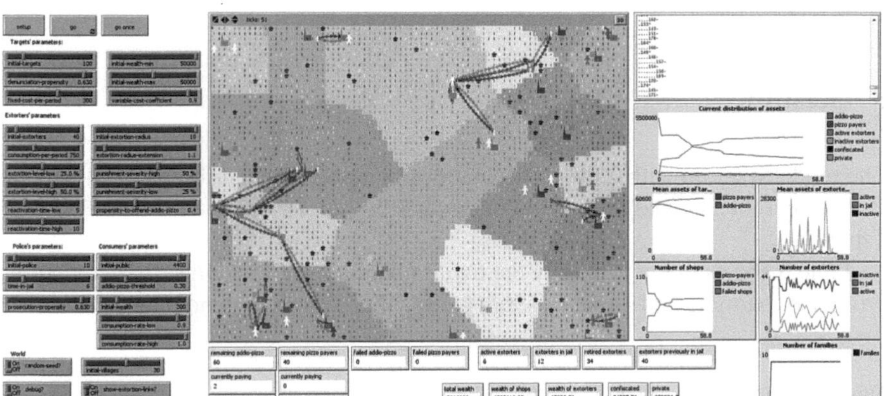

Fig. 1 The interface of the current extortion racket system model

[1]This is justified by the fact that it is not uncommon in Italy that negative effects of criminal activities are compensated by payments from a funds of confiscated mafia assets. See Legge 181/2008, Art. 2 comma 7, Gazetta Ufficiale n. 268, 15/11/2008, according to which at least one third of all confiscated assets are transferred to two funds, the solidarity fund for extortion victims and the solidarity fund for victims of other mafia type crimes.

[2]The code of the model can be found at http://ccl.northwestern.edu/netlogo/models/community/ARDERS.

villages whose centers are marked with church towers. Family links (green/orange) between extorters (person symbols, family heads in double size, in different colour marking the severity of their extortion and punishment demands, low-low: yellow, low-high: orange, high-low: red and high-high: magenta) as well as extortion links (blue/red) between extorters and shops (little houses, marked red to signalise "addio-pizzo" membership, blue otherwise, and black when they are temporarily closed) are shown as double arrows in different colours such that the development of the extortion racket system can be followed while the simulation is running. Consumers are marked with tiny person symbols (the darker the longer their distance to their current shop).

Several plots and monitors show the history and the exact current state, respectively, of several macro variables of the system. These are

Sums of the current and historical assets of shops (members or non-members of the "addio-pizzo" movement) and of the extorters (those active, those currently in jail and those inactive, i.e. whose assets are consumed),
means of the assets of shops and extorters,
assets of the consumers,
numbers of shops and extorters in their current roles,
the funds collecting confiscated money and distributing money to needy shops.

The hierarchy of the families of the extortion racket system can be seen in a text field (top right of Fig. 1).

3 Results

The model was run with 2,000 combinations of uniformly distributed input parameters (see Table 1), and the effects on several output variables were analysed (see Table 2).

Table 3 shows that it is mainly the two probabilities of the police to successfully prosecute extorters and—to a lesser degree—of the shops to denounce them which

Table 1 Input parameters considered for the analysis of the model

		Lower bound	Upper bound
Characteristic of	Input parameter	of the uniform distribution	
Extorters	Extortion-level-low	10 %	50 %
	Extortion-level-high	Extortion-level-low	75 %
	Punishment-severity-low	10 %	50 %
	Punishment-severity-high	Punishment-severity-low	75 %
Shops	Denunciation-propensity	0.1	0.7
Police	Prosecution-propensity	0.1	0.7
Consumers	Addio-pizzo-threshold	0.1	0.5

Table 2 Output variables considered for the analysis of the model

Parameter	Abbreviation for Table 3	Mean	Standard deviation
mean assets of all active shops ...			
... belonging to the "addio-pizzo" movement	m ass a-p shops	49,775	5,717
... not belonging to the "addio-pizzo" movement	m ass non a-p	37,509	3,404
the time when the "addio-pizzo" shops were richer than the others for the first time	t(m-ass-a-p>non-a-p)	12.42	10.878
mean assets of extorters ...			
... currently active	m-ass-ext-act	17,911	23,119
... currently in jail	m-ass-ext-jail	291	1,339
... currently inactive	m-ass-ext-inact	1,662	2,438
sum of all assets of all active shops ...			
... belonging to the "addio-pizzo" movement	$\sum$ ass a-p shops	2,942,390	352,667
... not belonging to the "addio-pizzo" movement	$\sum$ ass non a-p	1,499,454	495,534
sum of all assets of extorters	$\sum$ ass extorters	203,710	254,097
number of extorters who are head of a family	N families	5.97	1.688
max depth of a family tree	max-family-depth	11.35	2.341
size of largest family	max-family-size	13.28	5.648
number of all active shops...			
... belonging to the "addio-pizzo" movement	N a-p shops	60.13	11.330
... not belonging to the "addio-pizzo" movement	N non a-p	39.61	11.670
number of shops which are currently paying pizzo	N curr paying	3.82	3.273

determine the fate of this artificial economy: The former shares more than 75 % of its variance with the selection of output variables used here, and the prosecution propensity is positively correlated with the output variables denoting the success of the individual shops—whereas a high denunciation propensity increases the number of addio-pizzo shops and reduces the number of shops not participating in addio-pizzo, but the latter seem to profit from this effect as the correlation with the mean assets of the non-addio-pizzo shops is high and positive. The distribution of the overall economic success of the extorters is extremely skewed to the left which shows that for a majority of input parameter combinations the success of extorters is low, the main determinant for low extorter success being the prosecution propensity.

Table 3 shows clearly that it is mainly the denunciation and prosecuting propensities and—to a lesser extent—the addio-pizzo threshold which determine the main output variables: The denunciation propensity shares more than half of its variance ($R_\ell^2 = 0.530$) with all output variables, and in the case of the prosecution propensity the shared variance is even more than three quarters ($R_\ell^2 = 0.776$); the differences between the squared multiple regression coefficients with and without the other input parameters differ by small amounts in most cases (compare the last three columns of Fig. 3), i.e. the additional variance explained by the other four

Table 3 Correlation and squared multiple regression coefficients for the input parameters and output variables

Output variable	Input parameters									
	Extorter related				Shop	Police related	Consumer	All input parameters	Last three	two
	Extortion-level		Punishment-severity		Denuncia-tion-propensity	Prosecu-tion-propensity	Addio-pizzo -threshold			
	Low	High	Low	High						
	r	r	r	r	r	r	r	R^2	R^2	R^2
m ass a-p shops	−0.026	−0.023	−0.121	−0.135	−0.002	+0.793	+0.225	0.696	0.672	0.629
m ass non a-p	−0.151	−0.196	+0.012	+0.010	+0.307	+0.436	−0.416	0.524	0.473	0.285
t(m-ass-a-p>non-a-p)	−0.047	−0.077	+0.153	+0.145	+0.152	−0.480	−0.191	0.323	0.288	0.253
m-ass-ext-act	+0.011	+0.002	+0.181	+0.176	−0.048	−0.578	−0.016	0.618	0.580	0.366
m-ass-ext-jail	−0.025	−0.012	+0.053	+0.045	+0.002	−0.098	−0.044	0.015	0.011	0.010
m-ass-ext-inact	−0.011	−0.008	+0.123	+0.128	−0.044	−0.425	+0.028	0.207	0.184	0.182
$\sum$ ass a-p shops	−0.069	−0.042	−0.120	−0.148	+0.428	−0.358	+0.183	0.369	0.345	0.310
$\sum$ ass non a-p	−0.012	−0.034	−0.009	−0.008	−0.164	+0.787	−0.082	0.659	0.655	0.645
$\sum$ ass extorters	+0.001	+0.004	+0.163	+0.158	−0.045	−0.675	−0.031	0.496	0.458	0.458
N families	+0.016	−0.014	+0.034	−0.002	+0.288	+0.265	−0.006	0.157	0.154	0.154
Max-family-depth	+0.021	−0.014	+0.026	+0.014	−0.394	−0.420	−0.038	0.334	0.332	0.332
Max-famliy-size	−0.006	− + 015	−0.036	+0.003	−0.388	−0.392	+0.016	0.308	0.305	0.305
N a-p shops	−0.034	−0.015	+0.003	−0.006	+0.271	−0.757	−0.017	0.645	0.645	0.645
N non a-p	+0.032	+0.015	−0.011	−0.008	−0.267	+0.762	+0.025	0.652	0.651	0.651
N curr paying	−0.019	+0.002	−0.003	−0.007	−0.118	−0.735	+0.002	0.556	0.556	0.555
All output vars R^2	0.072	0.089	0.132	0.147	0.530	0.776	0.346			

variables is small and in some cases (number of shops belonging and not belonging to the addio-pizzo movement) minim.

The contribution of the denunciation probability alone is also small as compared to the one of the prosecution probability, but still large as compared to the four variables describing the strategies of the extorters. This input parameter adds to the linear regression of most of the output variables on the prosecution propensity only a modest amount.

The third most important input parameter is the addio-pizzo threshold, i.e. the probability with which consumers prefer an addio-pizzo shop over a shop which is in principle willing to pay pizzo, but it is mainly the mean assets of shops which is influenced by this parameter: addio-pizzo shops, of course, profit somewhat from this consumer preference ($r = 0.189$), the others suffer more intensively from the reluctance of consumers to support extortion indirectly ($r = -0.424$), but a closer inspection of the dependence of the non-addio-pizzo shop mean wealth on the addio-pizzo threshold of the consumers shows that the effect vanishes for thresholds beyond 0.1 (see the left plot in Fig. 2): For values between 0.0 and 0.1 of this parameter the slope of the regression curve is extremely steep—$\beta_a = -0.636$—(and the influence of the prosecution success seems to be low—$\beta_p = +0.208$—), whereas beyond 0.1 the slope of the regression curve is flat—$\beta_a = -0.137$—, and the influence of the prosecution success is high—$\beta_p = +0.637$—. A nonlinear approach to the dependence of the non-addio-pizzo shop mean wealth on the addio-pizzo threshold of the consumers raises the variance reduction from $R^2_\ell = 0.173$ to $R^2_{n\ell} = 0.385$. The dependence of the mean wealth of the addio-pizzo shops on the addio-pizzo threshold is similar but much less distinctive (see the right plot in Fig. 2).

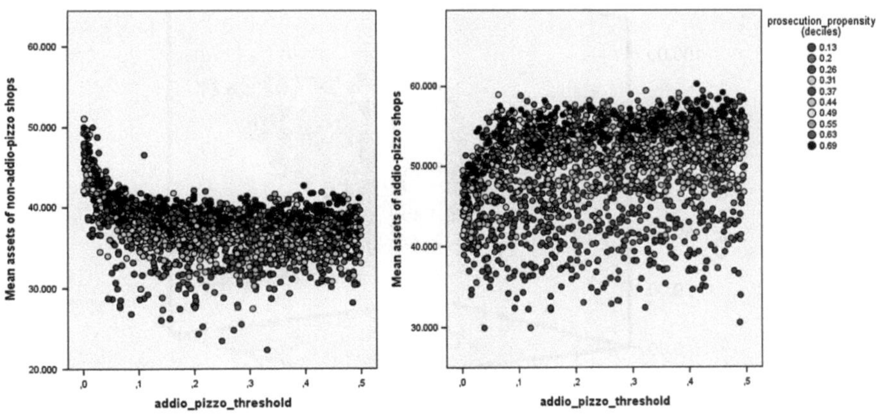

$y_1 = 36512.550 + \exp(-22.237x_a + 9.248)$ $\quad$ $y_2 = 50805.624 + \exp(-12.606x_a + 8.741)$

$R^2_{n\ell} = 0.385 \quad R^2_\ell = 0.173$ $\quad$ $R^2_{n\ell} = 0.069 \quad R^2_\ell = 0.051$

Fig. 2 Scattergrams showing the nonlinear dependence of the mean assets of remaining addio-pizzo and non-addio-pizzo shops after 50 simulation periods on the addio-pizzo threshold

As already mentioned, the four parameters describing the extortion and punishment strategies of the extorters do not contribute at all to the determination of the output variables; their share of the variance of all output variables together is between 7 and 15 %. Their additional variance reduction after regression on the three main input parameters falls between 0.001 and 0.051; thus they will not be discussed in detail as even a visual analysis of the respective scatterplots does not show any particularities.

Scattergrams of the correlations between the two propensity parameters and several output variables show a high degree of nonlinearity which can be seen in Fig. 4 and also some interaction between the denunciation and the prosecution propensity.

A nonlinear regression yields the following functions for the numbers of remaining non-addio-pizzo (y_1) and addio-pizzo (y_2) shops on the prosecution propensity (x_1) and the denunciation propensity (x_2) where the variance reduction is considerably higher than in the linear regression in Table 3—see also Fig. 3 which shows that only in a very small subset of the space spanned by the two propensities the number of non-addio-pizzo can be expected to be larger than the number of addio-pizzo shops—namely where the blue surface lies above the red surface (to be more precise, this is true for $x_1 > \exp(x_2)/2.2067$ as $x_1 = \exp(x_2)/2.2067$ is an approximate solution for $y_1 = y_2$ as defined in Eqs. (3) and (4)).

$$y_1 = 51.941 - \exp(-4.656x_1 + 4.083) \qquad R^2_{n\ell} = 0.649 \;\; R^2_{\ell} = 0.581 \qquad (1)$$

$$y_1 = 53.624 - \exp((-5.762 + 3.914x_2)x_1 + 4.065) \qquad R^2_{n\ell} = 0.713 \;\; R^2_{\ell} = 0.651 \qquad (2)$$

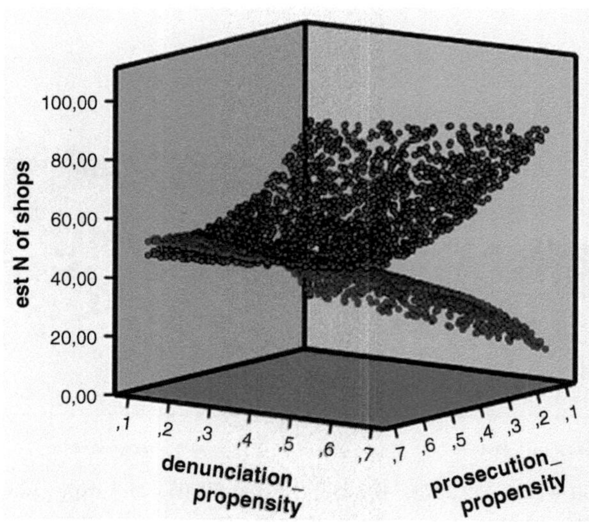

Fig. 3 Nonlinear regression (Eqs. (2) and (4)) of the numbers of remaining addio-pizzo (*red*) and non-addio-pizzo (*blue*) shops cn the denunciation and prosecution propensities (Color figure online)

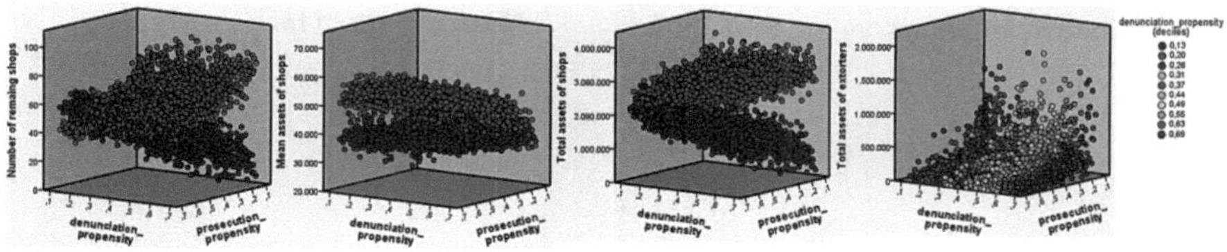

Fig. 4 Scattergrams showing the nonlinear dependence of the numbers of remaining addio-pizzo (*red*) and non-addio-pizzo (*blue*) shops, their mean assets, their total assets and the total assets of the extorters after 50 simulation periods on the denunciation and prosecution propensity (Color figure online)

$$y_2 = 47.433 + \exp(-4.2990x_1 + 4.012) \qquad R^2_{n\ell} = 0.632 \;\; R^2_{\ell} = 0.573 \quad (3)$$

$$y_2 = 45.500 + \exp((-5.274 + 3.644x_2)x_1 + 3.998) \qquad R^2_{n\ell} = 0.698 \;\; R^2_{\ell} = 0.645 \quad (4)$$

The interaction between denunciation propensity and prosecution propensity can be analysed with a three-dimensional view at the data particularly for the overall assets of the two types of shops.

The first plot in Fig. 4 (similar to Fig. 3, but showing all 2,000 runs instead of the two regression surfaces) shows the numbers of the two types of shops. Here more or less the opposite difference becomes visible: Most shops join the addio-pizzo movement for a high denunciation propensity (no surprise, this is at the same time the propensity to join) and a low prosecution propensity, whereas at the diagonally opposite end of the range of the two propensities a small majority of shops continues to surrender—these are the shops with a small propensity to denounce and to join the addio-pizzo movement in environments where the prosecution success is high (so why should they join?).

The second plot in Fig. 4 shows the mean assets of the two types of shops for different pairs of propensity values. Obviously the difference between the two means is minimal for a high denunciation propensity and a low prosecution propensity—in this extreme case denouncing is not useful, as the prosecution is very likely to fail, thus the fate of a shop joining the addio-pizzo movement is not better than the fate of a shop which surrenders to the extorters. At the other end of the diagonal, i.e. for a low denunciation propensity and a high prosecution propensity, the addio-pizzo shops are much better off—they are now rather safe from extortions, and consumers prefer them for their purchases.

The third plot in Fig. 4 combines the two effects visible in the first and third plot. It shows that the overall wealth of addio-pizzo shops is always higher than the one of the other shops (except for the extreme case where prosecution is effective and denunciation is rare. The higher the denunciation propensity and the lower the prosecution propensity, the more of the overall shop assets is with the addio-pizzo shops and the less is with the other shops—with a very big difference in the extreme case.

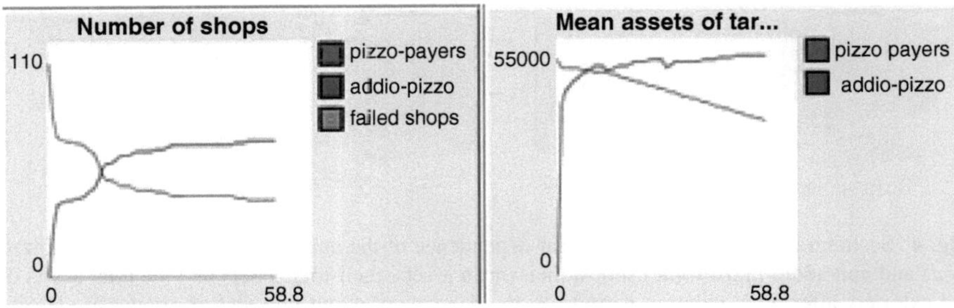

Fig. 5 History of the numbers and mean assets of the two groups of shops (*left*) and of the distribution of the total assets among the different owners for one typical simulation run

The fourth plot in Fig. 4 shows the overall wealth of the extorters in period 50 of each group of runs for pairs of propensity decile pairs. High prosecution propensities keep extorters in relative poverty, regardless of the propensity of the shops to denounce them. For lower prosecution propensities the slope of the surface showing the overall wealth of the extorters is steeper and steeper, and extorters are, of course, best off when both propensities are low.

So far we dealt only with the final states after 50 periods. What happens during typical simulation runs for different values of the main input parameters is shown in the final Figures 6 and 7 for the development of the criminals. The distribution of wealth between the two groups of shops, their mean wealth and their numbers develop rather smoothly and converge to their final values after about 10–20 periods, remaining more or less smooth or even constant afterwards—see Fig. 5 for just one out of the 2,000 runs, for other input parameter combinations the plots look quite similar though on different levels, and the intersection of the blue and red curves may come earlier or later.

Figure 6 shows that in most cases there is a wave of imprisonments very early in all of these simulation runs. After nearly every extorter spent some time in jail the future of the prosecutions and imprisonments develops differently: the higher the prosecution success the higher the number of extorters spending several periods in jail. Again the denunciation propensity plays a minor role—only for minor prosecution propensities a difference in the histories in Fig. 6 can be seen.

The mean economic success of individual extorters depends on both denunciation and prosecution propensity. For the smallest values of both relatively high amounts—up to more than 100,000 currency units can be seen for a short while, but even before and after this climax the average wealth of the active extorters continues to be around 70,000 c.u. With higher prosecution propensities the peaks are much lower, and in several runs they occur only at the onset of the simulations, and for the highest prosecution propensities the wealth of the extorters remains minim for the whole period (Fig. 7).

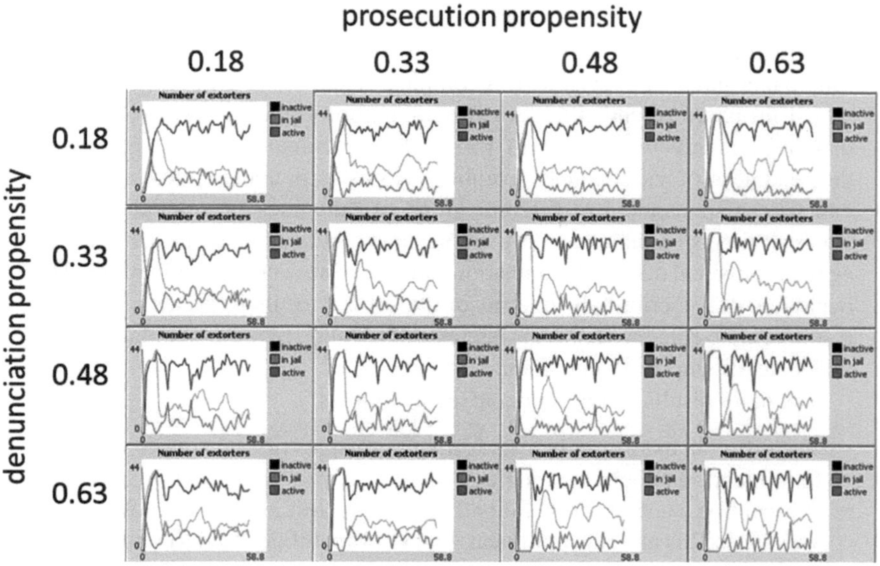

Fig. 6 Histories of the numbers of active and inactive extorters and those in jail for typical simulation suns with denunciation and prosecution propensities in the middle of their quartiles

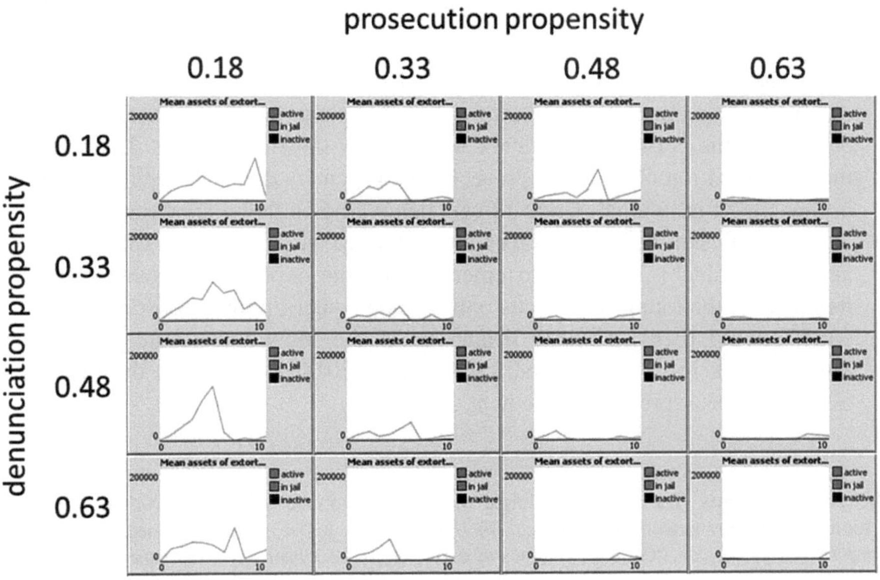

Fig. 7 Histories of the medium individual wealth of the extorters for typical simulation suns with denunciation and prosecution propensities in the middle of their quartiles

Conclusion

This paper gave an overview of a complex model of an artificial society in which firms are offended by an evolving extortion racket system. It shows that it is mainly the chance of prosecuting criminals (which depends on the readiness of victims and potential victims to denounce their extorters) which is responsible for different kinds of outcomes of simulation runs. Whether this is realistic can only be presumed, but can be empirically tested against statistical data of denunciations, prosecutions and condemnations. The behaviour of the criminals in terms of the severity of their threats plays only a subordinate role, certainly an interesting finding from the simulation runs which needs empirical validation.

The main results can be summarised as follows:

- The different extortion-and-punishment strategies of extorters do not seem to influence any output variables.
- Prosecution success probability has the greatest effect on most output variables. This effect is nonlinear, higher probabilities lead to saturation.
- Denunciation propensity has the second greatest effect, there is an interaction with the prosecution success probability.
- Consumers' willingness to differentiate between shops supporting and not supporting the addio-pizzo movement has the third largest effect, again this effect is nonlinear, and there is an interaction with the prosecution success probability.

Further extensions of the model will have to take into account that the denunciation propensity is not a constant for all shop owners but that it should depend on the personal experience acquired by the shop owners. Likewise, the simplification of a unique prosecution propensity could be replaced by a propensity of individual police officers depending on their experience and on characteristics of the denounced extorters. Group specific norms (Andrighetto and Castelfranchi 2013) could emerge from the experiences made by the members of the four groups—customers, shop owners, police and criminals—which would then guide the behaviour of the individual agents. This is already work in progress (Andrighetto et al. 2014) building on the results of a predecessor project (Conte et al. 2013).

Acknowledgements The research leading to these results has received funding from the European Union Seventh Framework Programme (FP7/2007–2013) under grant agreement no. 315874 (http://www.gloders.eu, "Global dynamics of extortion racket systems"). This paper reflects only the author's views, and the European Union is not liable for any use that may be made of the information contained therein. The author thanks his colleagues in this project for fruitful discussions over many years.

References

Andrighetto G, Castelfranchi, C (2013) Norm compliance: the prescriptive power of normative actions. Paradigmi 2:120–135

Andrighetto G, Lotzmann U, Nardin L, Neumann M (2014) Report on adaptations made to the EMIL simulator. The Global Dynamics of Extortion Racket Systems, Deliverable 3.1. http://www.gloders.eu/images/Deliverables/GLODERS_D3-1.pdf

Conte R, Andrighetto G, Campennì M (eds) (2013) Minding norms. Mechanisms and dynamics of social order in agent societies. Oxford University Press, Oxford

La Spina A (ed) (2008) I costi dell'illegalità. Mafia ed estorsioni in Sicilia. Il Mulino, Bologna

Pinotti P (2012) The economic costs of organised crime: evidence from southern Italy. Technical report, Banca d'Italia, Roma

Spina C (2009) Addiopizzo: can a label defeat the mafia? J Int Policy Solut 11:3–11

Troitzsch KG (2010) Communication and interpretation as means of interaction in human social systems. In: Martinás K, Matika D, Srbljinović A (eds) Complex societal dynamics: security challenges and opportunities. NATO science for peace and security series. IOS Press, Amsterdam, pp. 53–64

Wilensky U (1999) NetLogo. http://ccl.northwestern.edu/netlogo

Impacts on Stability of Interdependencies Between Markets in a Cobweb Model

Emma Jonson, Liv Lundberg, and Kristian Lindgren

1 Introduction

Financial markets are full of uncertainties that force economic actors to make decisions based on expectations. How the role of such expectations can stabilize or destabilize markets has been extensively, though not conclusively, studied with cobweb models. In its original form (Ezekiel 1938), the cobweb model explores the price dynamics of a single market when producers assume that current prices will hold. However, in reality economic actors are often active on, and affected by, several different markets.

A few models of interdependency between markets can be found in the cobweb literature. Waugh (1964) describes a linear model of the markets for feed and livestock where the output of any of the goods next year depends on the present prices of both of them. Conversely, the prices of feed and livestock are also functions of the actual output of both of the two goods. The markets are thus "linked" from both the supply and the demand sides. Just as the original cobweb model of a single market, this multidimensional linear model either converges, diverges or falls into a two-cyclic pattern.

Some studies have focused on the link on the demand side that exists between markets for substitute and complementary goods. Currie and Kubin (1995) explores the implications of ignoring seemingly negligible interdependencies between two such markets. (Here the supply side reacts only to the prices in the own markets.) The authors find that a partial model of either market is unsatisfactory when describing the price dynamics. Hommes and van Eekelen (1996), on the other hand,

E. Jonson (✉) • L. Lundberg • K. Lindgren
Department of Energy and Environment, Chalmers University of Technology, 41296 Göteborg, Sweden
e-mail: emma.jonson@chalmers.se; livl@chalmers.se; kristian.lindgren@chalmers.se

F. Amblard et al. (eds.), *Advances in Artificial Economics*, Lecture Notes in Economics and Mathematical Systems 676, DOI 10.1007/978-3-319-09578-3_16

question the strength of this conclusion. By including a small amount of noise, the partial model and the complete two-market model qualitatively and quantitatively yield the same results.

Supply side linkage has been studied by Dieci and Westerhoff (2010). Here, producers may choose to enter one out of two markets, which introduces nonlinearity into the model. The study shows that when markets are interlinked new dynamic features may emerge that are not found in any of the markets alone.

In this paper we present a model of cobweb markets that are interlinked on both the supply and demand sides, and apply it to land use competition between bioenergy and food production. The markets for these two types of commodities are inherently interlinked, since they both rely on land as a limited resource, and when global food prices increased drastically in 2006–2008 a debate emerged about the role of increased bioenergy production. In the policy discussion that ensued, various types of equilibrium modelling approaches were almost always used as a basis for conclusions (Persson 2014; Fargione et al. 2008; Searchinger et al. 2008; Hertel et al. 2010). However, methods including dynamic aspects may add important insights and act as a complement to the equilibrium models. One example of a question of policy interest is whether an increase in bioenergy demand causes higher instability of food prices.

In our model we start with three generic crops as an aggregated representation of the world's production of food and bioenergy. Equilibrium models used for bioenergy assessments, however, are typically more detailed. See for example Havlik et al. (2011) who present a model representing more than 30 of the most important crops. Including more crops and making models more detailed allows a closer representation of reality. It also tends to make the model larger and less transparent. Thus, there might be a tradeoff between detail and transparency. In this paper we analyse how our model is affected by increasing the number of crop types. We aim to answer the question of how much the system is destabilized by increasing the number of crops, and how much a "link" on the demand side (through substitutability of commodities) counteracts this destabilization. This question relates to the tradeoff between detail and transparency in models. If a model with few generic crop types gives similar results as a more detailed model with multiple crop types, this indicates that the system can be aggregated and described in a more simple and transparent form. Since models used for bioenergy assessments today are highly detailed, exploring the effects of this added detail, is relevant to model development as well as policy discussions.

In the research that we present, an agent-based model (ABM) is used to simulate the supply side of a global agricultural market. The agents are profit-maximizing farmers with heterogeneous production capacities that can choose between different crops to produce, thereby interlinking the markets for the crops. The consumer side of the model is represented by interlinked demand-functions that regulate the degree of substitutability between crop types. We find that supply side interdependency increases price fluctuations while demand side interdependency reduces fluctuations.

2 Agent Representation in Cobweb Models

In the original cobweb model, producers are homogeneous and their output is described by a simple linear supply function where quantities produced depend on expected prices. Different types of heterogeneity have since then been introduced. In Dieci and Westerhoff (2010) the agents are producers who choose one out of two markets to enter. An intensity of choice parameter decides how large fraction of producers that switch to the most profitable market in the preceding time step. The approach is similar to Brock and Hommes (1997), where agents pick one out of two prediction methods of future prices (one-step perfect foresight or naive expectations), based on realized net profits in the last period. These papers are both examples of cobweb models with heterogeneous agents (in choice of market to enter, and in choice of prediction method, respectively). In these cases the agents' aggregated actions are still simple enough to be described by a few equations. In other types of studies of the cobweb economy, ABMs are used to represent the supply side. This is the case in i.e. Anufriev et al. (2013) where learning to forecast future prices is modeled using a genetic algorithm. With ABMs, learning and adaptation of individual agents may be represented and information about the performance of decisions may spread in social networks. There is an increasing awareness that ABMs may be suitable for land use modelling since they can include heterogeneity in both land suitability and agent characteristics (Rounsevell et al. 2013).

In our research we have chosen a bottom-up approach where the supply side is represented by an ABM. This enables us to make farmer characteristics and the decision making processes explicit. The agents are heterogeneous in production capacity, which has a stabilizing effect that naturally exists in the global agricultural system.

3 Model Description

The framework for our model is a simple, conceptual agricultural land-use system. The agents are farmers that make decisions on how to use their land. The options are producing one out of N crop types or leaving the land idle. The model is intended to represent the global agricultural system in a highly stylised way, so the number of agents and land parcels is very large. For simplicity, in the model formulation, we assume a continuum of agents.

The landscape on which the model operates is defined by agricultural land of varying relative productivity, $Y \in [0, 1]$, but with no explicit geographic representation. The agents each own a piece of land of equal area, but different relative productivity, Y. Here we assume that Y varies linearly over the total agricultural land (currently about 5 Gha) which is an approximation derived from real world estimates found in Fisher et al. (2002). Land that is not used for agriculture today, such as forest areas etc., is not included in the model.

Crop production in the model is quantified in terms of energy content (*GJ*) since the main driver of the demand for food and bioenergy crops is the need of energy (for human metabolism and for the energy system, respectively). Each crop type i has a maximum potential yield of η_i $[GJ\ ha^{-1}]$, but only a fraction of this potential can be realized by the agents, depending on the relative productivity (Y) of the land that they own. The actual harvest (h) that crop i would yield on land belonging to agent a is $h_{a,i} = \eta_i Y_a$ $[GJ\ ha^{-1}]$. The agents are not allowed to choose the intensity of production (by adjusting amount of fertilizer etc.). The only choice that the agents make is what crop, if any, to produce.

Each crop type i is characterized by a harvest dependent cost β_i $[US\$\ GJ^{-1}]$ (pesticides, fertilizers, transportation to market etc.), and an area dependent cost α_i $[US\$\ ha^{-1}]$ (tillage and capital equipment etc.). The cost (c) for agent a of producing crop i is therefore

$$c_{i,a}(p_i) = \beta_i h_{a,i} + \alpha_i. \tag{1}$$

(This contrasts with the classical cobweb model where the supply curve is linear, implying that producers face quadratic cost functions.) At the market price p_i $[US\$\ GJ^{-1}]$, the profit ($\pi$) of producing crop i on the land belonging to agent a would be

$$\pi_{i,a}(p_i) = (p_i - \beta_i)h_{a,i} - \alpha_i. \tag{2}$$

Since the agents are heterogeneous in production capacity (defined by the relative productivity (Y) of the land they own), the profit that each crop can generate also varies among the agents. Different agents will find different crops most profitable. The choice of crops to produce is based on expected future profits. All crops are assumed to be harvested at the same time and put on a global, ideal market where their prices are determined by an inverse isoelastic demand function,

$$p_i(q_i) = p_{i,0}\left(\frac{q_i}{q_{i,0}}\right)^{\frac{1}{\varepsilon_i}}, \tag{3}$$

where q_i is the total amount produced by all agents of crop i, $\varepsilon_i < 0$ is the elasticity of demand and $p_{i,0}$ and $q_{i,0}$ are constants representing the prices and quantities in equilibrium. (In the classical cobweb model the demand curve is linear. However, for such essential commodities as food crops, we assume that quantities demanded never drop to zero, and that prices always are positive.)

Time runs in discrete time steps and each crop takes one period to produce. Only a fraction, γ, of the producers are allowed to change crop at each time step. The rest of the agents produce the same crop as last time step. This kind of inertia could be motivated since a switch of crops is associated with new investments in knowledge and production capital. A farmer may be unwilling to make these investments if he has just recently already made a switch in production. Inertia may also be a result of farmers being influenced by tradition, esthetics, contracts etc. The crop a land owner would like to sow is the one that maximizes profits if prices would stay the same until the time of harvest. There is no randomness in the current implementation of the model.

3.1 Application to a Food and Bioenergy Scenario

We consider three main categories of crops: intensively produced edible type and forage crops (IP), extensively produced permanent pasture and forage crops (EP) and bioenergy crops (BE), adopted from Bryngelsson and Lindgren (2012, 2013). BE crops include crops that could be grown under relatively commercial conditions for bioenergy production, such as perennial lignocellulosic crops. Parameters defining the supply and demand functions for the crops can be seen in Table 1. The equilibrium quantity (q_0) of BE corresponds roughly to the present global supply of modern bioenergy. However, an important share of the current bioenergy feedstock consists of by-products from the agricultural and animal sectors, residues from industry and municipal solid waste, whereas BE in our model represents bioenergy crops produced on dedicated plantations. Our scenario therefore corresponds to a situation where the bioenergy market is somewhat expanded compared to today. The fraction of farmers (γ) allowed to change crop in each time step is 10 %.

3.2 Demand Side Interdependency

In the basic case we have one generic crop representing each category (intensive production, IP, extensive production EP and bioenergy BE). We now divide each

Table 1 Crop specific parameter values used in all calculations

	η [GJ ha^{-1}] maximum potential yield	α [US\$ ha^{-1}] area dependent cost	β [US\$ GJ^{-1}] harvest dependent cost	q_0 [EJ] equilibrium demand	p_0 [US\$ GJ^{-1}] equilibrium price	ε own-price elasticity
IP	90	500	4	60	12	−0.5
EP	70	50	1	95	3.55	−1
BE	250	300	3	10	6	−0.5

See Bryngelsson and Lindgren (2013) for sources of data
IP intensive production, *EP* extensive production, *BE* bioenergy

category into n crop types that are substitute goods (see Fig. 1). These sub-crops could be thought to represent for example wheat, maize, rice etc. (in the case of IP crops), grass-legume hay, whole-maize, forage-vegetables etc. (in the case of EP crops) and eucalyptus, miscanthus and willow etc. (in the case of BE crops). However, in the model we keep the sub-crops generic and do not adjust the parameters in order to correspond to any particular real-world crop.

In the following part of the paper, q_{ij} denotes the quantity produced of crop (i, j), where $i \in \{IP, EP, BE\}$ and $j \in \{1, 2, \ldots, n\}$. The total quantity produced of crops from category i is still denoted q_i, so that $q_i = \sum_j q_{ij}$. The equilibrium quantity of each crop is denoted $q_{ij,0}$. The sum of the equilibrium quantities within each crop category is the same as in the basic case ($\sum_j q_{ij,0} = q_{i,0}$). The equilibrium price $p_{i,0}$, the cost parameters β_i, and α_i and the potential yield parameter η_i all remain the same within the crop category as in the basic case. The prices are determined by

$$p_{ij} = p_{i,0}\left(\frac{q_{ij}}{q_{ij,Ref}}\right)^{\frac{1}{\varepsilon_i}} \quad (4)$$

Interdependency on the demand side between markets is regulated by the parameter $q_{ij,Ref}$ which we define as:

$$q_{ij,Ref} = \omega \frac{q_{ij}}{q_i} q_{i,0} + (1 - \omega) q_{ij,0}. \quad (5)$$

with $\omega \in [0, 1]$ defining the level of substitutability between commodities. When $\omega = 0$ we have $q_{ij,Ref} = q_{ij,0}$ and the demand for crop (i, j) becomes completely independent of the quantities produced of the other crops in the same category (and thereby also by their prices). When $\omega = 1$ the crops are perfect substitutes and Eq. (4) becomes identical to Eq. (3). In this case the price per GJ is the same for

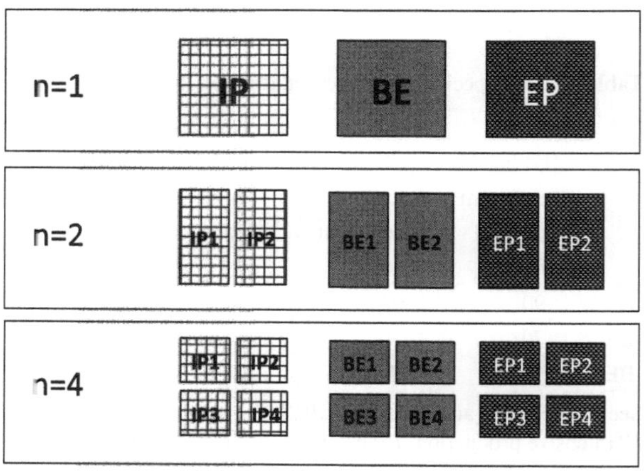

Fig. 1 Each crop category is divided into n crop types. The level of substitutability of the crops within a crop category is defined by the parameter ω, introduced in Eq. (5). The crop categories are intensive production (IP), extensive production (EP) and bioenergy (BE)

all crops within the same category i. Equation (5) is chosen as a simple, schematic, representation of demand side interdependency.

4 Results

In order to explore the effects of supply and demand side interdependency on price fluctuations we use numerical simulations. Ten percent of the agents are allowed to change production per time step (with each time step representing a new year).

The first simulation experiment illustrates the degree of stability of the isolated markets for IP, EP and BE. One single generic crop type represents each crop category. Here, the agents may not choose which crop to produce, only if they should produce their designated crop or not. Time series of quantities produced and prices of the three crop categories are shown in Fig. 2a, d, respectively. We can see that the markets for IP and EP are in isolation stable, whereas the BE market is highly unstable.

Figure 2b, e show the case when the three crop types are "linked" on the supply side by letting the agents choose which of the crops to produce. Quantities and prices of all three crops now exhibit semi-periodic fluctuations over the years. We can deduce that the instability of the system is a result of BE being present as a land use option. The instability of the BE market is transferred to the IP and EP prices when the three markets are interlinked. The effect of interdependency on the supply side is presented in more detail and discussed in Lundberg et al. (2014).

The next simulation shows the effect of introducing more crop types. When each generic crop type is divided into four sub-categories (thus making the total number of crops 12) that are non-substitutable ($\omega = 0$), the system becomes much more unstable and chaotic than the basic case. Figure 2c shows the total quantities produced of each generic crop category (the sum of the quantities in the four sub-categories) per year. The total demand for each crop category is the same as in the basic case, and is split equally between the sub-categories. Aggregating the quantities in the figure facilitates comparisons with the basic case (shown in Fig. 2b). We can see that the total quantities produced in each crop category is much more volatile in the $n = 4$ case than in the basic case ($n = 1$). Since the prices can not be aggregated, Fig. 2f shows the individual prices of all 12 crop types, but with crops of the same generic category labelled with the same colour.

The increased volatility of the model can be explained by the fact that any absolute change in production of a crop from one year to another now means a larger relative change. (With more crops included in the model, the quantities produced in equilibrium of each crop is smaller.)

We now study the effect of demand side interdependency by varying the degree of substitutability between the sub-categories, ω [see Eq. (5)], from 0 (non-substitutable) to 1 (perfect substitutes). We compare the basic case ($n = 1$) with the case of two and then four crop types per category. We see that the degree of price fluctuations is strongly related to the level of substitutability between the sub-

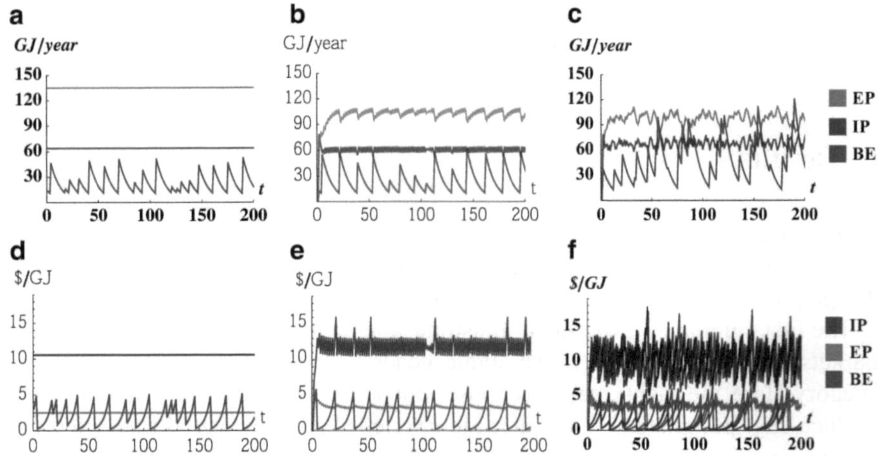

Fig. 2 Fluctuations in prices and quantities produced for different levels of supply side linkage. (**a**) and (**d**) show the three markets in isolation. In (**b**) and (**e**) the three markets are linked on the supply side by letting agents choose which crop to produce. (**c**) and (**f**) show time series of quantities and prices, when each generic crop category is divided into four sub categories (thus making the total number of crops 12) that are unlinked (non-substitutable) on the demand side. In (**c**) the total quantities of each generic crop type is shown (the sum of the quantities in the four sub-categories), while (**f**) shows the individual prices of all 12 crop types (even though crops that are of the same generic category is labelled with the same colour). What can be observed in the figures is that the instability of the BE market is transferred to the IP and EP markets when they are linked on the supply side. The system becomes even more unstable and chaotic when the generic crop types are divided into sub categories that are not interlinked on the demand side. (**a**) $n = 1$, Isolated markets. (**b**) $n = 1$. (**c**) $n = 4$. (**d**) $n = 1$, Isolated markets. (**e**) $n = 1$. (**f**) $n = 4$

categories. With perfect substitutability the number of sub-categories is basically irrelevant for system stability. This can be seen in Fig. 3a, where $\omega = 1$ (perfect substitutability) and the percentage standard deviation of prices is unchanged when the number of sub-categories is varied ($n = 1, 2, 4$). If the sub-categories are only partly substitutable as in Fig. 3b, where $\omega = 0.5$, an increased number of sub-categories is correlated with increased standard deviation of prices. The percentage standard deviation increases even more with increased number of sub-categories when they are non substitutable ($\omega = 0$) as in Fig. 3c. The price fluctuations are smaller with a demand side linkage, since the prices now affect each other. Any extreme price is moderated by the prices of the other crops in the same category.

Conclusions and Discussion

The recent cobweb literature recognizes that markets may not always be studied satisfactorily in isolation since there are links to other markets. We

(continued)

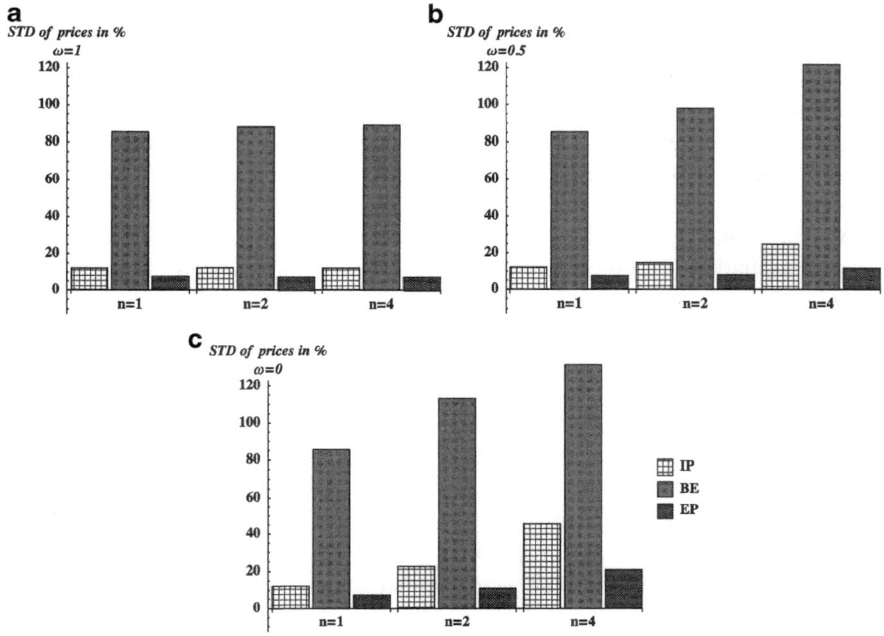

Fig. 3 Standard deviation of prices, in percent, for IP, BE and EP with each generic crop category divided into 1, 2 or 4 sub crops. In (**a**) the sub crops within each crop category are perfect substitutes ($\omega = 1$). In this case the percentage standard deviation is unaffected by the number of sub crops within the categories. When the sub crops are only partly substitutes, $\omega = 0.5$, or independent on the demand side, $\omega = 0$, (**b**) and (**c**), the standard deviation is increased with number of sub crops. The basic case where $n = 1$ is included as a reference case in (**b**) and (**c**). (**a**) Standard deviation of prices, with $\omega = 1$. (**b**) Standard deviation of prices, with $\omega = 0.5$. (**c**) Standard deviation of prices, with $\omega = 0$

present a cobweb model of interdependent markets on both the supply and demand sides and apply it to a food and bioenergy framework, with a stylized representation of global variation in land quality. The supply side of the model is implemented as an agent based model where farmers in discrete time steps make decisions on how to use their land.

In a first setting crops are aggregated into crop categories, with one generic crop representing each category. The markets are interdependent on the supply side through the limited availability of land. When a bioenergy crop is added as a land use option to the model, the system becomes highly unstable. The bioenergy market is in isolation unstable and by linking it to the food markets, instability is transferred, causing fluctuations in food crop prices.

(continued)

In a second setting we divide each crop category into multiple sub crops. We find that increasing the number of crops leads to higher instability in prices. However, this instability can be moderated by introducing a demand side interdependency; a cross-elasticity of demand between the crops in the same category. When the crops are perfect substitutes, the level of volatility becomes comparable to the first setting with only one crop per category. We show that the two kinds of interdependencies have opposing effects. Linking markets on the supply side transfers instabilities within the system and may cause price fluctuations in previously stable markets (Lundberg et al. 2014). Market interdependency on the demand side, on the other hand, has a stabilizing effect.

In the model that we present there are three categories of crops. *Intensively produced crops (IP)* include all food crops that are consumed directly by humans, and all intensive production of feed crops used for raising of swine, poultry and cattle. *Extensive production (EP)* includes permanent grasslands for grazing and for low-intensive production of feed crops used for ruminants. *Bioenergy crops (BE)* are crops of perennial lignocellulosic type. In this study, demand side interdependency is studied only by considering cross-price elasticity of demand between crops in the same category. Demand for crops in different categories is assumed to be completely independent. This approach can of course be questioned. It is not unreasonable to assume that IP and EP are substitute goods to a certain degree. If the price of IP rises, the demand for ruminant meat (and thereby implicitly the demand for EP) may increase, and vice versa.

Another question is how the own-price elasticity of demand is affected by the disaggregation of crops within each crop category. In the model, the elasticity of demand for the crops is the same within each crop category, regardless of how many crops we have. A reasonable assumption is that the aggregate demand for all food crops is less elastic than the demand for individual crops. This problem is taken care of by introducing interdependency described by Eq. (5). If the production of a certain sub-crop is very small, its price would be very high, were interdependency not included. As the interdependency parameter ω grows larger, the price of any sub-crop is more and more governed by the total production of all crops within the category.

The division of crops within the generic categories is done in a stylized way so that crops within each category have identical cost characteristics. In future work a diversification of crop types could be introduced in order to create an agent-based counterpart to large equilibrium models with extensive sets of crops. The demand side could also be replaced by an agent based model, which would provide a more explicit representation of consumer preferences.

In this study we have shown that the level of price fluctuations in a cobweb model of land use depends on the supply and demand side representation.

(continued)

In a dynamic model designed to study price fluctuations it is important to take into account that the results will depend on the level of aggregation. The higher number of crops the model has, the larger impact does any "links" on the supply and demand sides have on the model results.

Acknowledgements Financial support from the Swedish Energy Agency and from the EU-FP7 project MatheMACS is gratefully acknowledged.

References

Anufriev M, Hommes C, Makarewicz T (2013) Learning-to-forecast with genetic algorithms. Technical report, working paper

Brock WA, Hommes CH (1997) A rational route to randomness. Econ J Econ Soc 65(5):1059–1095

Bryngelsson DK, Lindgren K (2012) A conceptual partial equilibrium model of global agricultural land-use. Working paper. Available at http://publications.lib.chalmers.se/records/fulltext/local_164501.pdf

Bryngelsson DK, Lindgren K (2013) Why large-scale bioenergy production on marginal land is unfeasible: a conceptual partial equilibrium analysis. Energy Policy 55:454–466

Currie M, Kubin I (1995) Non-linearities and partial analysis. Econ Lett 49(1):27–31

Dieci R, Westerhoff F (2010) Interacting cobweb markets. J Econ Behav Organ 75(3):461–481

Ezekiel M (1938) The Cobweb theorem. Q J Econ 52:255–280

Fargione J, Hill J, Tilman D, Polasky S, Hawthorne P (2008) Land clearing and the biofuel carbon debt. Science 319(5867):1235–1238

Fisher G, van Velthuizen H, Shah M, Nachtergaele F (2002) Global agro-ecological assessment for agriculture in the 21st century: methodology and results. International Institute for Applied Systems Analysis (IIASA) and Food and Agriculture Organization of the United Nations (FAO), Laxenburg/Rome

Havlik P, Schneider UA, Schmid E, Böttcher H, Fritz S, Skalskỳ R, Aoki K, Cara SD, Kindermann G, Kraxner F, et al. (2011) Global land-use implications of first and second generation biofuel targets. Energy Policy 39(10):5690–5702

Hertel T, Golub A, Jones A, O'Hare M, Plevin R, Kammen D (2010) Effects of us maize ethanol on global land use and greenhouse gas emissions: estimating market-mediated responses. BioScience 60(3):223–231

Hommes C, van Eekelen A (1996) Partial equilibrium analysis in a noisy chaotic market. Econ Lett 53(3):275–282

Lundberg L, Jonson E, Lindgren K, Bryngelsson D, Verendel V (2014) A cobweb model of land-use competition between food and bioenergy crops (submitted)

Persson UM (2014) The impact of biofuel demand on agricultural commodity prices: a systematic review. Accepted for publication in WIREs Energy and Environment

Rounsevell M, Arneth A, Brown D, de Noblet-Ducoudr N, Ellis E, Finnigan J, Galvin K, Grigg N, Harman I, Lennox J, Magliocca Ns, Parker D, ONeill B, Verburg P, Young O (2013) Incorporating human behaviour and decision making processes in land use and climate system models. GLP Report No. 7

Searchinger T, Heimlich R, Houghton RA, Dong F, Elobeid A, Fabiosa J, Tokgoz S, Hayes D, Yu T (2008) Use of U.S. croplands for biofuels increases greenhouse gases through emissions from land-use change. Science 319:1238–1240

Waugh FV (1964) Cobweb models. J Farm Econ 46(4):732–750

References

Detecting Key Variables in System Dynamics Modelling by Using Social Network Metrics

J. Barranquero, M. Chica, O. Cordón, and S. Damas

1 Introduction

Strategic management requires special economic and marketing planning, offering the ability to comprehend and anticipate the effects of complex dynamic interactions between a firm and their business environments and stakeholders. This is therefore a complex social system that requires understanding emergent patterns and their systemic implications (Bonabeau 2002; Dickson et al. 2001). A concrete example of this strategic problem is the brand value management, where decision makers must consider the outcome of their investments to make a sustainable and differential advantage relative to their competitors (Aaker 1996).

Building a business dynamic model that lays out the critical resources of the scenario and the key relationships between them offers a competitive gain for decision makers. This kind of modelling also provides a way to carry out simulations and understand the effects of the different policies. Among other methodologies, system dynamics (SD) (Forrester 1961; Sterman 2000) presents a theoretical framework with a set of tools and techniques for developing mathematical models of complex systems for social and economic scenarios.

The SD methodology is particularly useful in systems with many interrelated variables, where relevant data to build the system is not always available. SD offers

J. Barranquero • M. Chica (✉) • S. Damas
European Centre for Soft Computing, Mieres (Asturias), Spain
e-mail: jose.barranquero@softcomputing.es; manuel.chica@softcomputing.es; sergio.damas@softcomputing.es

O. Cordón
European Centre for Soft Computing, Mieres (Asturias), Spain

DECSAI and CITIC-UGR, University of Granada, 18071 Granada, Spain
e-mail: oscar.cordon@softcomputing.es; ocordon@decsai.ugr.es

F. Amblard et al. (eds.), *Advances in Artificial Economics*, Lecture Notes in Economics and Mathematical Systems 676, DOI 10.1007/978-3-319-09578-3_17

the opportunity to simulate a problem by investigating its results and behaviour, making the framework useful for policy testing, what-if scenarios, or policy optimization.

The set of applications of SD is enormous (O'Regan and Moles 2006; Winz et al. 2009). Besides, it has played an important role for a systemic view of management issues (Warren 2005) and marketing applications. The application of SD for brand management assists marketing experts in understanding how different factors affect the value of a particular brand, how costumers react to a brand in terms of loyalty and equity, the influence of email marketing campaigns, or the effects of implementing innovation policies in organizational policies (Mukherjee and Roy 2006; Richardson and Otto 2008).

However, it is sometimes difficult to identify key variables in dense or large problems modelled by SD. These key variables are those able to generate significant changes in the whole system. This descriptive information of the system is vital for modellers since they can apply strategic actions over those variables (in a direct or indirect way) and focus their what-if scenarios. The identification of these key variables is also useful for understanding the dynamics of the model and for validation purposes, given that key variables might constitute an additional boundary adequacy and structure verification test for the model (Oliva 2003; Qudrat-Ullah 2012).

The main focus of the current paper is how to detect which variables of a SD model constitute the set of key variables. Our proposal is to first compute a quality metric for every variable of the graph structure of the model. These values indicate the importance role of each variable with respect to the whole structure of the model. Then, we rank model's variables according to this metric, suggesting those that yield better values.

Hence, our proposed quality metric is founded on network-based properties of the model structure and is therefore applied on the whole SD graph. The computation of the metric is based in turn on the *scope* and *closeness* of an agent within a social network, which are well known metrics in social network analysis (SNA) (Carrington et al. 2010; de Nooy et al. 2005; Oliveira and Gama 2012).

We have modelled and simulated a TV show brand management problem to validate the application of our key variable detection algorithm. This systemic abstraction is based on an existing work that analyzes the Indian version of "Who wants to be a millionaire" (Mukherjee and Roy 2006). We have followed Vester's sensitivity model (Vester 1988, 2007) to shape the system dynamics and structure. This SD methodology is convenient for sustainable processes and enables analysts to simplify the real world complexity into a simulation and consensus system. After applying the algorithm and extracting the key variables of the model structure we run different simulations to compare the global impact of injecting strategic actions just over top-ranked key variables.

The rest of the paper is structured as follows. In Sect. 2 we study the background and describe the SD modelling and social networks metrics of our proposal. Then, Sect. 3 contains the analysis of results and simulation graphs of the key variables detection for the TV show case. Finally, we present some concluding remarks in section Concluding Remarks.

2 Methods

2.1 System Dynamics for Modelling Complex Marketing Systems

There are different methodologies and tools for system dynamics modelling. For our work we follows the sensitivity model proposed by Vester (1988, 2007) which offers a semi-quantitative SD modelling tool based on systems thinking and fuzzy logic (Zadeh 1975). It has been applied to different fields of research, environmental and risk management, and tourism (Huang et al. 2009; Meyer-Cech and Berger 2009; Schianetz and Kavanagh 2008). The main advantages of this approach are the ease of use and the employ of feedback analysis as the core component of the modelling process.

There are nine steps in sensitivity modelling. These include system description, set of variables, criteria matrix, impact matrix, systemic role, effect system, partial scenarios, simulation and cybernetic evaluation. They can be categorized into three phases (see Fig. 1). The first phase begins with a general system description and the

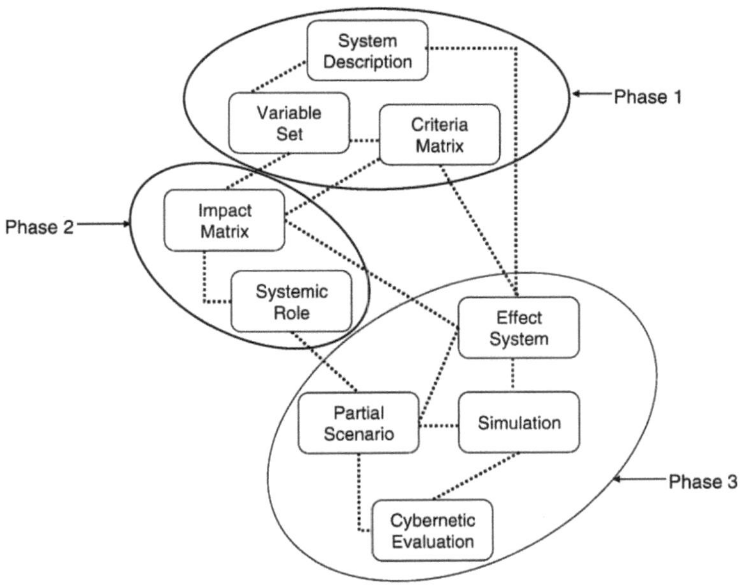

Fig. 1 The main steps of the sensitivity model methodology (Huang et al. 2009)

identification of influential factors for system development. In the second phase there is an analysis of the magnitude of the cause-effect relations between the variables and to identify their functional roles in the system (model structure). Simulation and cybernetic evaluation of phase 3 are based on the framework of the effect system and the positive and negative feedback relationships. The partial scenario of the focus issue can be simulated to observe the dynamics and interrelationships among variables.

One of the major components of this SD methodology is the feedback loops that embody the information feedback structure of the system. Feedbacks are defined by the effects among the variables of the system. An effect between two variables can be direct or inverse. The effects, in conjunction with the variables, form the graph structure of the system. See Fig. 2 for a structure example of the case study of this work.

The simulation results arise from this interaction among feedback loops. Feedbacks are of two types: mitigating (an initial change in one variable of the loop will finally change the variable in the opposite direction, balancing the initial change) and reinforcing (where the initial change will be reinforced through the feedback process).

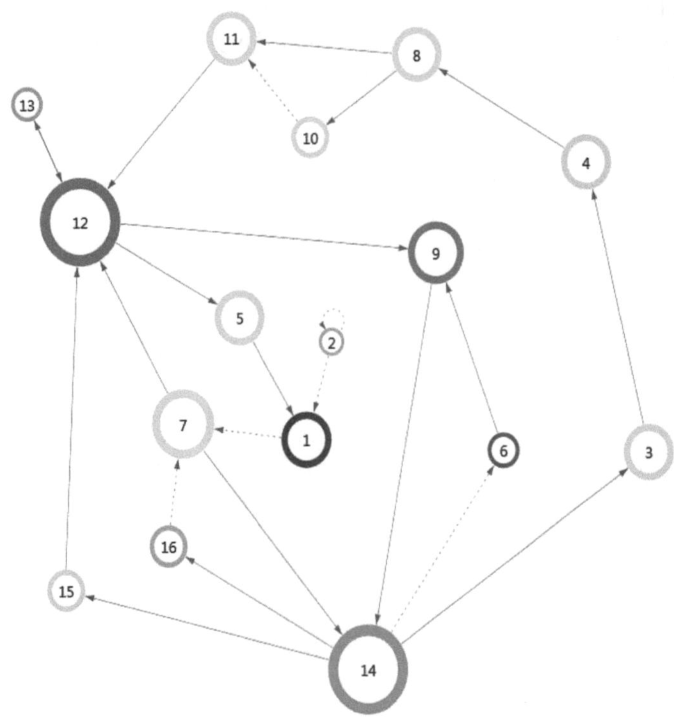

Fig. 2 System model structure for TV show case study

For this work, we extend the original SD methodology to tackle economical and marketing problems as ours. Therefore, the modeller can also define the details of effects (delay, intensity, and values of change) and variables (initial and optimum value, limits, blocked status, and randomness). In order to run simulations the model allows the definition of the temporal horizon and different strategies to be applied to some action variables to optimize the state variables of interest.

2.2 Use of Social Network Analysis for Key Variable Detection

We define the key variables of a system as those inherently relevant due to their interconnection with others variables of that system. We consider that a key variable is not required to be a good lever for defining specific actions, even when it plays an important role in the development of the system.

Therefore, instead of focusing on specific simulation conditions for key variable detection, we consider that it is more interesting to emphasize the intrinsic structure of the network, which in fact represents the system dynamics through its effects. This shall generate a more stable and general key variable set, suitable for a broader range of configurations.

2.2.1 Social Network Centrality Metrics

Centrality metrics are typically applied in SNA, making intensive use of statistical graph-based measures (Carrington et al. 2010; de Nooy et al. 2005; Oliveira and Gama 2012). SNA distinguish two levels of analysis: individual units (variables, actors, etc.) and whole network. In this paper, we focus on the former, given the definition of key variables that we want to address, clearly oriented to explore the role of each individual. Whole network indicators may be also helpful for obtaining more condensed knowledge, although we do not cover them in detail for this work due to length constraints.

Therefore individual SNA metrics can be adapted for our purposes, given that they share a common objective of identifying key actors in a network. Globally, these metrics are considered as a measure of *centrality* or *prestige*, in which the most common (Bonacich 1987; Freeman 1979) are:

- **Degree or valency**: analyzes the immediate neighbourhood of each node and is computed as the number of edges of the node. For directed graphs it is divided into in-degree (input prestige or support) and out-degree (output prestige or influence). Its biggest drawback is that it is a local measurement that does not reflect the global structure of the network.
- **Betweenness**: measures the relevance in terms of the number of shortest paths that go though the node, although it can be also computed for edges. Nodes (or edges) with high betweenness are supposed to interact heavily in information

flow and diffusion between communities (i.e., to play a strong brokerage in the information diffusion process).

- **Closeness**: evaluates how fast a given individual can reach the whole network, defined as the average length of all shortest paths with origin on the node. Its main problem is that it is not defined for those cases where there exist pairs of nodes that are not connected by any path.
- **Eigenvector centrality**: a re-elaboration of degree, which takes into account the quality of first order connections. It is computed after assigning a relative score to each node, measuring their connectivity with other well-connected nodes.

2.2.2 The Proposed Quality Metric

The reviewed metrics lack to take into account the number of reachable nodes, which can be defined as its *scope*. The scope measures how many variables can be reached directly or by transitivity. For instance, the standard closeness, defined in the previous section, could be highly skewed. This is because variables with low scope may show optimal values of closeness, while being poorly connected. In other words, it is considered a local measure.

Therefore, we have adapted the original definition of closeness in order to take into account the scope and the delay of each effect (weight of the edge in our model). This proposal allows measuring both concepts jointly, considering on the one hand the elasticity and penalizing on the other hand the absence of a path between nodes.

The new metric is termed *elastic distance* (ED) and is computed as the average of the shortest weighted distances from the source variable to all other variables. The distances to non-reachable ones are fixed with a sufficiently large value in order to embody information about the scope of the variable under study, defined as supreme-distance constant M:

$$ED(i) = \frac{1}{n} \sum_{j=1}^{n} d(i, j), \forall i \neq j, \tag{1}$$

where $d(i, j) = M$ when there is not any path between nodes i and j.

3 Experimentation Results for a TV Show Case Study

In this section we apply our proposed metric to a SD model adapted from the original model of the Indian "Who wants to be a millionaire" TV show (Mukherjee and Roy 2006). The graph of the model is presented in Fig. 2. Node colours depend on current value of corresponding variables, while node diameter is defined in terms of number of feedbacks in which the variable participates. The name, metric value

Table 1 Metric values obtained for variables defined in TV show case study

ID	Name	Metric	Description
12	Interest level	5.93	Perceived interest by show viewers. Equivalent to brand equity defined by Aaker (1996)
2	Innovation	6.20	Action variable that allows influencing the system by simulating novelties like special editions
7	Brand loyalty	6.33	State variable that measures actual viewers loyalty. One of the measures of brand value defined by Aaker (1996)
14	Actual viewers	6.47	Measures the success of the show in terms of total number of viewers. Critical state variable
13	Host popularity	6.80	Current popularity of the person driving the show. Reinforcing feedback cycle with interest level
1	Repetitiveness	6.87	Measures the degree of repetitiveness of the show. It is directly influenced by episodes rate and innovation
9	Probability of joining	7.27	Probability of viewers joining in. It depends on interest level and potential viewers
5	Episodes rate	7.40	Number of episodes per time unit. Too many episodes affect negatively to repetitiveness and brand loyalty
6	Potential viewers	8.07	Available viewers that do not follow the show currently. Inversely related with actual viewers
10	Minimal promotion level	8.73	Minimum spending for promoting the show. This variable tends to increase over time
11	Promotion effectiveness	8.87	Effectiveness of investment in promoting the show. It acts directly over interest level
15	Channel popularity	9.47	Similar to host popularity, though less influenced by show's interest level
16	Competition	9.53	Amount of competing shows. Influenced by actual viewers and influences brand loyalty
8	Promotion expenditure	10.73	Amount of money spent on promotional campaigns. Influenced by advertisement revenue
3	Advertisement rate	13.60	Advertisement benefit per time unit. It is heavily influenced by actual viewers
4	Advertisement revenue	13.67	Advertisement incomes per time unit. Closely related with advertisement rate

and description of each variable is detailed in Table 1 (variable list is sorted in terms of elastic distance).

3.1 Results of Key Variable Detection Algorithm

The two highest ranked variables are *interest level* (5.93) and *innovation* (6.20), while variables with the lowest rank are *advertisement rate* (13.60) and *advertisement revenue* (13.67). Given that our proposed metric is a weighted distance, this ranking reflects the speed and scope of propagation of changes for each variable.

Interest level is ranked first because it has many outgoing paths, spreading changes over the rest of system variables very quickly. The case of *innovation* is quite similar, because it allows a rapid access to core system variables, including interest level.

Advertisement rate and *advertisement revenue* obtain bad ranks because they are placed at the beginning of a long path, which implies a slower propagation over the whole network.

3.2 Simulation Results Using the Set of Key Variables

In order to validate our proposed algorithm for key variable detection, we present two simulation scenarios in Fig. 3. The idea is to test if there exists a significant difference between acting over top ranked versus bottom ranked variables.

The system is initially configured to be in a relatively steady state, avoiding strong trends that could clutter the interpretation of simulation results, with respect to a simulation baseline. The simulation engine evolves all variables of the system with a range that goes from 0 to 100, representing abstract values without specific representation units.

We simulate two alternative strategies with two actions each. The first strategy (Fig. 3, left graph) acts over the two lowest ranked variables, modifying their values to optimum. We refer to this scenario as *strategy L*. The second one (Fig. 3, right graph) applies the same change to optimum values over the two highest ranked variables, referred as *strategy H*. Finally we also simulate the system without any action, in order to measure the % of change of each variable with respect to *baseline simulation*. All simulations are performed over a 1 year period.

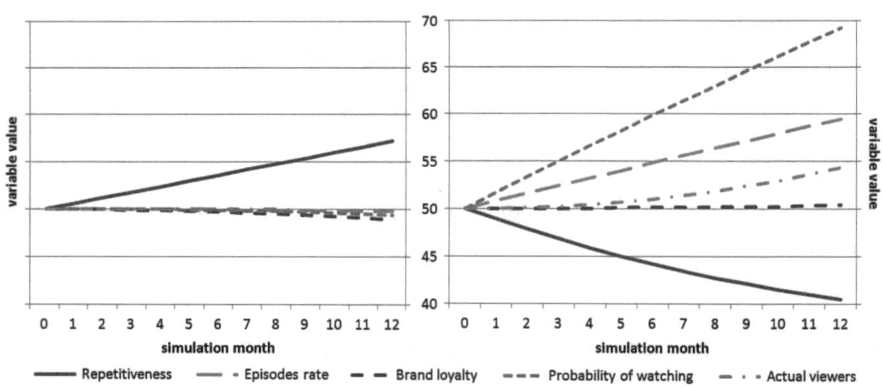

Fig. 3 Evolution of the system when directly acting over the lowest ranked variables (strategy L, *left*) and over the top key variables (strategy H, *right*)

We analyze the obtained simulation results excluding actioned variables from both strategies and measuring absolute changes with respect to baseline simulation. We also define two basic metrics, *total absolute distance* (TAD):

$$TAD = \sum_{i=1}^{n} |v_i - v_i'|, \tag{2}$$

and *mean absolute distance* (MAD):

$$MAD = \frac{1}{n} \sum_{i=1}^{n} |v_i - v_i'|; \tag{3}$$

where v_i and v_i' represent the final value of variable i for simulated baseline and strategy respectively.

Strategy L only affects 25 % of variables (3/12), obtaining a TAD of 15.6 and a MAD of 1.3. On the other hand, strategy H produces changes in 67 % of variables (8/12), with a TAD of 71.5 and a MAD of 6.

The highest change produced by strategy L is achieved over *promotion expenditure*, with a variation of 14.8 (out of 15.6 TAD). This is because it is the closest variable on the main path that starts on actioned variables. In this simple case it may seem obvious that the strategy is worthless, but at least the experiment succeeds in validating that the algorithm is effectively detecting the less critical variables.

Strategy H achieves an outstanding improvement with respect to both the baseline and strategy L. The strongest changes are produced over *probability of watching* (19.8), *host popularity* (18.9), *repetitiveness* (16.7), and *episodes rate* (9.6). All of these variables are in turn on the middle upper part of the ranking of key variables, producing a snow ball effect that shall be more significant for longer simulations.

Concluding Remarks

A key variable detection algorithm to be applied over the structure of a SD model was presented in this work. A quality metric is calculated for each variable of the model to quantify its importance for changing the evolution of the system. This metric is an extension of closeness, a widespread SNA measure, that we use for ranking the set of variables of the model structure.

The results of the proposed detection algorithm can be an effective validation test for the designed model. The set of key variables point out which variables are prevailing in terms of the model structure. Hence, if the set does not fit with the intended idea of the system, probably the design of the model may require a revision.

We tackled a brand management problem for a TV show, modelling it by SD and applying the key variable detection algorithm. Results showed how

(continued)

variables *interest level* and *innovation* were the most important of the model yielding a metric value of 5.93 and 6.20, respectively. Strategic actions were applied to these variables to present the impact in the simulation results with respect to a baseline simulation. Our experiments showed that acting over these two key variables have a remarkable effect over system variables, with an average improvement of 6 points (MAD metric) over system variables in 1 year. The same experiment but applied over the two lowest ranked variables only produced an average deviation of 1.3 points (MAD metric).

Some future works arise from this contribution: (1) propose and evaluate other quality metrics such as *Local Clustering Coefficient* (Watts and Strogatz 1998) or algorithms like *Pagerank* (Brin and Page 1998; Easley and Kleinberg 2010); (2) include additional measurements for providing complementary information about the model structure; and (3) develop an optimization engine for defining appropriate strategic actions in order to maximize profitability in what-if scenarios.

Acknowledgements This work has been supported by RØD Brand Consultants under ZIO project and Ministerio de Economía y Competitividad under SOCOVIFI2 (TIN2012-38525-C02-01 and TIN2012-38525-C02-02) projects.

References

Aaker DA (1996) Measuring brand equity across products and markets. Calif Manag Rev 38(3):102–120

Bonabeau E (2002) Predicting the unpredictable. Harv Bus Rev 80(3):109–116

Bonacich P (1987) Power and centrality: a family of measures. Am J Sociol 92(5):1170–1182

Brin S, Page L (1998) The anatomy of a large-scale hypertextual Web search engine. Comput Netw ISDN Syst 30(1):107–117

Carrington PJ, Scott J, Wasserman S (2010) Models and methods in social network analysis. Cambridge University Press, Cambridge, UK

de Nooy W, Mrvar A, Batagelj V (2005) Exploratory social network analysis with Pajek. Cambridge University Press, Cambridge, UK

Dickson PR, Farris PW, Verbeke WJ (2001) Dynamic strategic thinking. J Acad Mark Sci 29(3):216–237

Easley D, Kleinberg J (2010) Networks, crowds, and markets: reasoning about a highly connected world. Cambridge University Press, Cambridge, UK

Forrester JW (1961) Industrial dynamics. Cambridge, US

Freeman LC (1979) Centrality in social networks conceptual clarification. Social Netw 1(3):215–239

Huang SL, Yeh CT, Budd WW, Chen LL (2009) A sensitivity model (SM) approach to analyze urban development in Taiwan based on sustainability indicators. Environ Impact Assess Rev 29(2):116–125

Meyer-Cech K, Berger H (2009) Spatial impact of a factory outlet center in a small Austrian community - the case study of Leoville. disP Plan Rev 45(176):19–30

Mukherjee A, Roy R (2006) A system dynamic model of management of a television game show. J Model Manag 1(2):95–115
Oliva R (2003) Model calibration as a testing strategy for system dynamics models. Eur J Oper Res 151(3):552–568
Oliveira M, Gama J (2012) An overview of social network analysis. Wiley Interdiscip Rev Data Min Knowl Discov 2(2):99–115
O'Regan B, Moles R (2006) Using system dynamics to model the interaction between environmental and economic factors in the mining industry. J Clean Prod 14(8):689–707
Qudrat-Ullah H (2012) On the validation of system dynamics type simulation models. Telecommun Syst 51(2–3):159–166
Richardson GP, Otto P (2008) Applications of system dynamics in marketing: Editorial. J Bus Res 61(11):1099–1101
Schianetz K, Kavanagh L (2008) Sustainability indicators for tourism destinations: a complex adaptive systems approach using systemic indicator systems. J Sustain Tourism 16(6):601–628
Sterman J (2000) Business dynamics: systems thinking and modeling for a complex world. McGraw-Hill, New York
Vester F (1988) The biocybernetic approach as a basis for planning our environment. Syst Pract 1(4):399–413
Vester F (2007) The art of interconnected thinking: tools and concepts for a new approach to tackling complexity. MCB Publishing House, Munich, Germany
Warren K (2005) Improving strategic management with the fundamental principles of system dynamics. Syst Dyn Rev 21(4):329–350
Watts DJ, Strogatz SH (1998) Collective dynamics of 'small-world' networks. Nature 393(6684):440–442
Winz I, Brierley G, Trowsdale S (2009) The use of system dynamics simulation in water resources management. Water Resour Manag 23(7):1301–1323
Zadeh LA (1975) Fuzzy logic and approximate reasoning. Synthese 30(3–4):407–428

Trade-In Programs in the Context of Technological Innovation with Herding

Paolo Pellizzari, Elena Sartori, and Marco Tolotti

1 Introduction

In the Summer of 2013, Apple launched a new updating campaign in the US to entice old iPhone 4 users to switch to the new iPhone 5 (see Rampell 2013). The updating cost was really tempting (customers received a gift card up to \$250 to purchase the new version). Why is Apple *underpricing* its brand new technology? What's the rationale behind this pricing campaign?

In this paper we use an agent-based model (ABM) to study the revenues that can be achieved by a monopolist, which will be referred to as the *firm* in what follows, in a setup with competition between technologies issued at subsequent dates and volatile consumers. In the last decades several ABMs dealing with innovation and technological change have been proposed (see Adner and Levinthal 2001; Dawid 2006 for a recent review). These papers mainly focus on different patterns of investment strategies for production, heterogeneity of firms with respect to knowledge accumulation, abilities and levels of expertise. The goal is to try to forecast market reactions, considering most of the aspects involved in the decision to invest on incremental innovations, with minor extensions to existing products, or radical ones, with the idea of opening new markets.

P. Pellizzari
Department of Economics, Ca' Foscari University, Venice, Italy
e-mail: paolop@unive.it

E. Sartori
Dept. of Mathematics, University of Padua
e-mail: esartori@math.unipd.it

M. Tolotti (✉)
Department of Management, Ca' Foscari University, Venice, Italy
e-mail: tolotti@unive.it

F. Amblard et al. (eds.), *Advances in Artificial Economics*, Lecture Notes in Economics and Mathematical Systems 676, DOI 10.1007/978-3-319-09578-3_18

In Deffuant et al. (2005), an ABM to study diffusion of innovation is proposed. In particular, the authors discuss about the rationality of potential adopters and concentrate on their behavioral aspects, in order to show how a market share for a new technology or product forms and evolves over time.

The efficacy of ABMs in the context of diffusion of innovation has been also pointed out in Kiesling et al. (2012). As argued by the authors, although very useful to represent heterogeneous populations, beliefs and network structures, ABMs usually do not take into account competition among technologies. On the first hand, we extend the literature considering two competitive technologies and, on the second hand, we offer an analysis of how market shares are affected by the herding intensity, whose changes can proxy the entry of competitors in the business.

In our model technologies are exogenously given. What we focus on is the problem a firm has to face when deciding to launch new technologies on the market. Indeed, it has to set prices in order to maximize its own revenues. To do so, it forecasts demand and, accordingly, decides the optimal pricing strategy. On the other hand, potential adopters form their expectations about the diffusion of the technology. As a matter of fact, multiple equilibria for the market shares may arise. Using an ABM, we can analyze statistical properties of revenues, prices and market shares and show how, depending on the values of the parameters, completely different outcomes may emerge. Customers are price sensitive, get utility from imitation (herding) and take into account private signals about their willingness to pay (Brock and Durlauf 2001). The two main drivers of our model are, thus, the strength of the imitative behavior, which is anecdotally of great import in the technological market, and the prices set by the firm. We discuss the level of adoption rates at equilibrium as well as the determination of the firm's pricing schedule to maximize revenues.

In Nadal et al. (2005), a one-generation model is studied and a unique price is determined by maximizing the revenues from sales of a unique technology. Motivated by the Apple case described above, we generalize previous work and propose a two-phase model, where the competition among two generations of products makes the picture more involved and realistic. Indeed, facing a two-period game, the firm needs to forecast the emerging adoption rate at the end of the first round, in order to optimally allocate prices for the second period.

We show that, as somewhat expected, the optimal revenues are strongly influenced by the level of imitation on the market. Interestingly and paradoxically, for certain values of the parameters, the firm should optimally give away the update for free, maintaining its market share and boosting revenues.

The paper is organized as follows. In Sect. 2 we describe the decision process of potential adopters (demand) and the optimization scheme of the firm (supply). Section 3 is devoted to the specification of the ABM model. Section 4 addresses the main findings and section Discussion and Conclusion concludes.

2 Demand and Supply: Innovators, Updaters and Leapfroggers

Two sides have to be modeled: a large market of possible adopters (demand) and a firm that sets the prices. We start by the former.

2.1 *The Demand Side*

We consider a population of N possible adopters dealing with the problem of deciding whether to buy an existing technology. In particular, we consider two technologies, T_1 and T_2, issued at two subsequent periods. Note that the two technologies are not available to the market at the same time: T_1 can be bought only in period 1, whereas technology T_2 can be bought during period 2. Moreover, when deciding about the adoption for the first technology, the adopters are not aware of the second generation (or do not take it into account) and can only decide about T_1. An equilibrium analysis of the decision concerning a single technology has been discussed in the literature, see, for instance Nadal et al. (2005).

During the first period, the agents simultaneously choose whether to adopt T_1 or not. For each agent, we define the actions $a_1(i) \in \{0, 1\}$, for $i = 1, \ldots, N$, where

$$a_1(i) = \begin{cases} 1 & \text{if agent } i \text{ decides to adopt } T_1, \\ 0 & \text{otherwise.} \end{cases}$$

Without loss of generality, we normalize utility from non-adoption to 0, and set utility from adoption as follows

$$U_1(i) = -p_1 + q\,x + \varepsilon_1(i), \tag{1}$$

where $p_1 > 0$ is the price of T_1; $q > 0$ is the parameter measuring social utility coming from imitation; x is the expected *market share* for T_1; $(\varepsilon_1(i))_{i=1,\ldots,N}$ are i.i.d. random variables with distribution η. The utility of purchase depends on three summands: an individual noisy term $\varepsilon_1(i)$, a positive externality coming from the share of other users who adopt the technology and a negative effect due to the cost of adoption. Each agent compares his own utility of adoption and non-adoption, taking into account costs and social/private benefits. For N fixed, at least one Nash equilibrium in pure strategies $a_1^*(i)$ exists (see Dai Pra et al. 2013 for details). It can be characterized in terms of an equilibrium market share $x^{(N)} = \frac{1}{N}\sum_{i=1}^{N} a_1^*(i)$.

Moreover, when the number of possible adopters goes to infinity, any equilibrium market share x is proved to be a solution of the following implicit equation

$$x = 1 - \eta(p_1 - q\,x). \tag{2}$$

If η is unimodal, (2) can have one or three solutions depending on the values of p_1 and q. The problem described above belongs to a class of models dating back to the celebrated *riot* model by Granovetter (see Granovetter 1978), with x representing the fraction of people taking part in a riot. In the presence of herding, a quarrel could morph and escalate into a full blown uprising but, more importantly, the paper makes clear that (equilibrium) results may be extremely sensitive to small changes in the distribution of agents' preferences. This insight is helpful in describing some of the most prominent findings of our model and will be touched upon later.

In period 2 a new technology T_2 is issued and T_1 can no longer be adopted; besides, a T_1 owner can upgrade to T_2, if desired. Note that, at this stage, there are two different groups of agents: the first consists of *leapfroggers* (agents who did not adopt T_1 in the first period), the second of *innovators* (agents owning T_1). Depending on the group, agents face different utilities. Concerning leapfroggers, we set utility of non-adoption at zero, whereas utility from adoption (denoted by U_{02}) is

$$U_{02}(i) = -p_{2A} + q\,y + \varepsilon_2(i), \tag{3}$$

where p_{2A} is the cost for buying T_2; y is the expected market share for T_2 and $(\varepsilon_2(i))_{i=1,\dots,N}$ are i.i.d. terms with distribution η representing random terms related to the second technology. Note that y is now defined as $y = y_A + y_U$, where y_A and y_U denote, respectively, the proportion of new adopters and updaters in period 2.

Concerning innovators, they have to decide between keeping T_1 or updating to T_2. For these groups of agents, the utilities to be compared are

$$\begin{aligned} U_{11}(i) &= q\,(x - y_U) + \varepsilon_1(i), \\ U_{12}(i) &= -p_{2U} + q\,y + \varepsilon_2(i), \end{aligned} \tag{4}$$

where U_{11} denotes the utility from maintaining T_1, U_{12} the utility from updating, p_{2U} is the updating cost and x is the market share for T_1. The term $x - y_U$ is the proportion of the *aficionados*, who prefer to hold T_1. This is the market share one considers, when evaluating the social utility of being locked in T_1.

Similarly to period 1, once the distribution of the noise terms and p_{2A}, p_{2U} are fixed, the emergent market share can be computed by solving an implicit system of two equations. In particular, when the number of agents tends to infinity, the equilibrium market shares y_A and y_U can be formally characterized by the implicit system

$$\begin{cases} y_A = (1-x)\left[1 - \eta\big(p_{2A} - q\,y\big)\right] \\ y_U = x\left[1 - \tilde{\eta}\big(p_{2U} - q\,(y - (x - y_U))\big)\right], \end{cases} \tag{5}$$

where $\tilde{\eta}$ denotes the distribution of the random variable $(\varepsilon_2(i) - \varepsilon_1(i))$ conditional on the event $\{a_1(i) = 1\}$. If (2) admits a unique solution x and η is unimodal, then (5) has one or three solutions. To the best of our knowledge, there is no closed-form for the solution of the system (5). Moreover, we are interested in computing market shares when the number of possible adopters is large, but finite. In Sect. 3 we will show how to take advantage of an ABM to compute the emergent market shares.

2.2 *The Supply Side*

The firm chooses prices to maximize revenues, namely the vector $\mathbf{p} = (p_1, p_{2A}, p_{2U})$. Once prices have been set, agents form their demand and market shares emerge. We denote them by the vector $\mathbf{m} = (x, y_A, y_U)$, where we drop the dependence on $\mathbf{p}$ for simplicity. Total revenues are

$$\Pi_q(\mathbf{p}, \mathbf{m}) = p_1 x + p_{2A} y_A + p_{2U} y_U = \mathbf{p} \cdot \mathbf{m}', \tag{6}$$

where q is assumed to be exogenous in our model. The strength of imitation q has a very important role in the optimal pricing decision, as shown in the next sections. Notice, moreover, that the objective function (6) depicts a situation in which the revenues in both periods are important as well as interrelated. Indeed, p_1 has a double role: it (explicitly) determines the revenues of the first period and (implicitly) shapes the picture of the market in the second period. Indeed, one of the outcomes of the first period stage is the identification of innovators and leapfroggers; this distinction clearly affects the second wave of revenues.

3 The Agent-Based Model

We now provide details about the construction of the ABM.

Initial set-up

- We consider all values of q in the grid $\{2.0, 2.1, \ldots, 3.9, 4\}$. As already said, q will remain the unique exogenously fixed parameter in the model. All the other quantities (prices and shares) will be hereafter determined optimally.
- We fix a three-dimensional grid of values, where triplets of prices $\mathbf{p} = (p_1, p_{2A}, p_{2U})$ are chosen. In the present simulation, we have[1] $p_1 \in \{1.00, 1.05, \ldots, 1.95, 2.00\}$, $p_{2A} \in \{0.50, 0.55, \ldots, 1.45, 1.50\}$ and $p_{2U} \in \{-0.5, -0.4, \ldots, 1.4, 1.5\}$.

[1]Ranges for p_1, p_{2A} and p_{2U} have been selected after some preliminary explorations of the location of optimal solutions.

- We consider a finite and large population of agents ($N = 1{,}000$).
- We simulate $M = 50$ different stories for each of the price configurations (triplets) within the grid.

Demand: First period

- Agent i receives a private signal $\varepsilon_1(i)$: his/her personal view on T_1. Moreover, all the state variables are fixed at zero (nobody owns the first technology).
- The N agents choose their actions according to utility as in (1). The outcome[2] is a market share x^N.

Demand: Second period

- Agent i receives a private signal $\varepsilon_2(i)$: his/her personal view on T_2.
- Agents are divided into two groups depending on their action at period 1: leapfroggers or innovators.
- Leapfroggers choose their actions according to (3), whereas innovators rely on (4). The outcome[3] is a market share $y^N = y_A^N + y_U^N$.

Supply

- We evaluate the M revenues for each of the price configurations (triplets) in the grid of values.
- For any given q, we select $\mathbf{p}^*(q) = \arg\max_{\mathbf{p}} \Pi_q(\mathbf{p}, \mathbf{m}(\mathbf{p}))$ as the vector at which the median revenues (over the M simulations) are maximized.
- Then we form, for each q, the set P_q of price triplets, whose M simulated revenues are not significantly different from the revenues obtained with $\mathbf{p}^*(q)$ (using a Wilcoxon test with 5 % significance level). The elements, i.e., prices, of P_q generate revenues that cannot be statistically distinguished from the optimal ones.

Final outcome

- The set of optimal prices (strategies) P_q, together with the emerging market shares and optimal revenues, both in batches of M items, for each q in the grid.

The reader may wonder about the role of the set P_q: recall that prices are discretized and revenues are noisily estimated using simulations. As a result, there may be several different optimal price triplets that should be considered as indistinguishable by the firm as they produce the same revenues. While typically P_q is a "ball" centered at $\mathbf{p}^*(q)$, there are values of q for which this set has a non trivial

---

[2]The prevailing market share is obtained simulating repeated adoptions till convergence is reached: in the first round a fraction of x_1 agents will adopt according to (1) even in the absence of other adopters; in the second round more agents will join based on the current x_1 and the total fraction rises to x_2; the process continues till the adopters' share stabilizes at some t, i.e., $x_t = x_{t+1}$. The prevailing market share is then defined to be $x^N = x_t$.

[3]The prevailing market shares are obtained as in period 1, using fictitious rounds of adoptions till convergence is reached.

structure, suggesting that entirely different pricing schedules, supporting diverse market shares, are nevertheless equivalent in terms of revenues. This strategic multiplicity can be appreciated only when P_q, as opposed to $\mathbf{p}^*(q)$ alone, is analyzed. For a detailed analysis on this aspect, see point 4 in Sect. 4.

4 Results

In this section we outline the main findings and the results derived by the ABM procedure.

1. *Revenues increase with q*. The higher is q, the higher are the expected revenues, see Fig. 1, left panel. Imitation as measured by q plays a key role in determining the level of the revenues that are almost linearly increasing. It is interesting to decompose the total revenues in the three components due to adoption in period 1, adoption in period 2 and updating (from T_1 to T_2) in period 2. Clearly, adoption in the first period brings a major share of revenues, especially when q is high. The income due to adoption in the second period (dashed line) is decreasing with q reflecting, among other things, a mechanical effect: if the share of innovators increases, the group of leapfroggers shrinks. The revenues coming from updaters (dash-dotted line) are small and relatively flat. We will shortly show that this nearly constant stream of revenues is indeed attributable to a rich dynamics connecting the share of adopters and the price of update.
2. *The share of updaters peaks at a critical q*. Figure 1 on the right depicts the (optimal) market shares as functions of q. The share of innovators appears

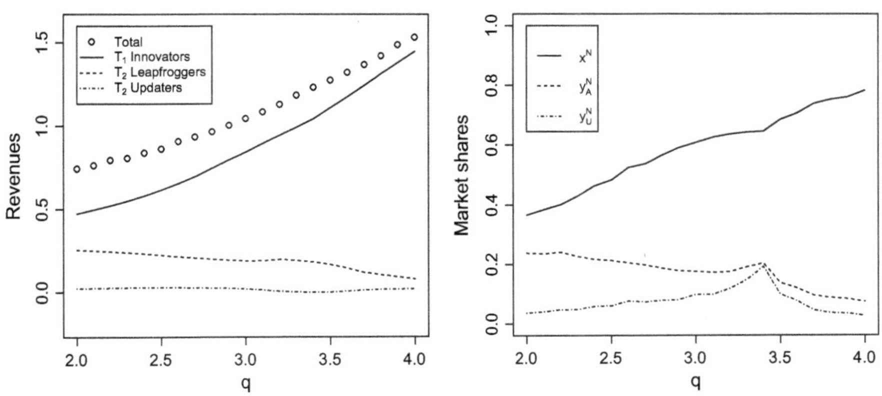

Fig. 1 *Left panel*: total revenues Π_q as a function of q (*circles*). The *three lines* show the revenues of the first period (*solid line*) and of the adoption due to leapfroggers (*dashed line*) and updaters (*dash-dotted line*), respectively. *Right panel*: shares of innovators (adopters in period 1), leapfroggers and updaters (adopting or updating in period 2). The three fractions are shown with *solid*, *dashed* and *dash-dotted lines*, respectively

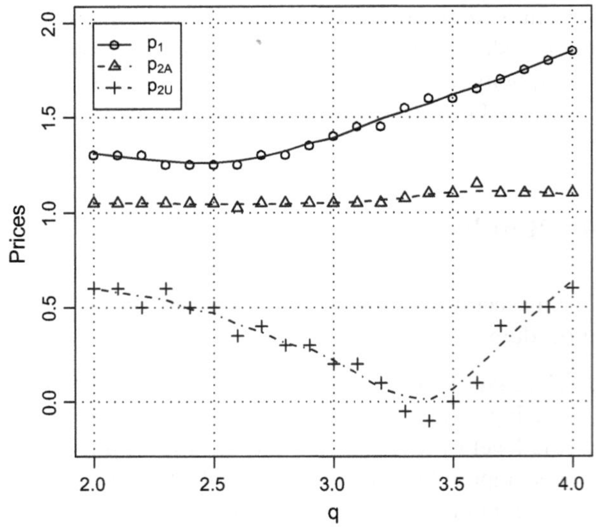

Fig. 2 Prices for different q: median optimal p_1, p_{2A} and p_{2U} in the set P_q are shown with *circles*, *triangles* and *crosses*, respectively. Simulation results are smoothed with *solid*, *dashed* and *dash-dotted lines* for added clarity

to steadily increase, whereas the fraction of leapfroggers decreases (with a blip around $q \approx 3.4$). Interestingly, the share of updaters peaks at the same critical q (dotted line) and reaches 20 % of the customers' base. Intuitively, the increase in the number of updaters helps in pushing more leapfroggers to adopt, thus explaining the temporary deviation from the declining trend of their share (dashed line).

3. *The updating price is very low for some values of (critical) q*. Figure 2 depicts the median optimal prices in the set P_q. The price of adoption p_1 in the first period is nearly constant for $q \leq 2.7$ and increases for higher values of q. The price of delayed adoption by leapfroggers is virtually constant across all levels of herding.

   The most unexpected and captivating effect is related to the updating price, which is a U-shaped function of q. As a matter of fact, updates are very cheap, to say the least, in a range of values of q roughly centered at $q \approx 3.4$, which we previously defined as "critical". Simulations indeed suggest that it may be paradoxically optimal to *subsidize* updates, being the raw estimates negative in several instances. If we conservatively look at the smoothed dotted line, still we get the insight that updates should optimally be favored with aggressive pricing and massive discounts.

   Low updating prices are clearly responsible for the surge in the share of updaters visible at $q \approx 3.4$; in turn, this rise in the number of customers employing T_2 fosters further (delayed) adoptions by agents who did not adopt T_1 in the first period.

   This provides a strong rationale for the Apple campaign that, actually, offered owners of iPhone 4 a substantial discount to induce them to switch to the new iPhone 5. When externality is sufficiently high (approximately in the interval

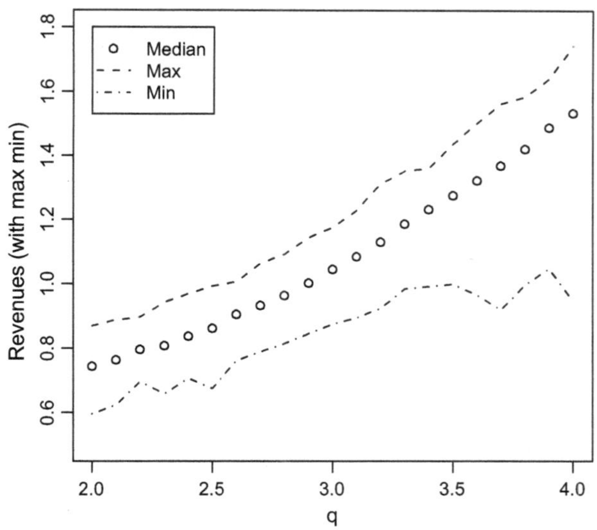

Fig. 3 Median revenues (*circles*), with their maximum and minimum levels computed considering all the triplets in the set P_q

$q \in [3.2, 3.6]$), it becomes optimal to *give the update for free*. This, in essence, is necessary to boost adoptions of leapfroggers, thus maintaining a high market share. Observe that, in the same range of the parameter q, the price p_{2A} of adopting T_2 stays constant, suggesting that it is preferable to inflate adoptions by enticing old users to upgrade rather then by diminishing the price of the new technology for newcomers. When q is very high ($q \geq 3.7$), it is optimal to raise again the updating price, p_{2U}: the huge externality will sustain new adoptions, regardless of the price.

4. *Variability of revenues increase at the critical q.* Figure 3 shows the total revenues already represented in Fig. 1, left panel (dotted line), but reporting also the maximum/minimum values obtained among the simulations relative to prices in P_q (dashed and dash-dotted lines, respectively). Patently, the dispersion of possible outcomes increases when q exceeds 3.4 and revenues fan out. In particular, for such values of q, low values of revenues become significantly frequent. Such variability may be produced by the presence of multiple equilibria in (2) or in the system (5). However, even in the lack of analytical multiplicity of the asymptotic model, sampling fluctuations can give rise to considerable variability of outcomes, as pointed out in Granovetter (1978).

   The contour lines of the simulated joint densities of x^N and y_U^N for $q = 2.8$ and $q = 3.4$ are shown in Fig. 4. The left panel represents market shares for a non-critical $q = 2.8$. Typical outcomes nicely crowd in an annular neighborhood of $(0.55, 0.10)$ and variability in the shares in this case is only attributable to sample fluctuations of the N idiosyncratic shocks of the agents. In the right panel relative to the critical $q = 3.4$, two broad situations are likely to materialize. Selecting an optimal price for that q can produce a continuum of market shares (x^N, y_U^N) clustered around $(0.7, 0.1)$ *and* $(0.6, 0.4)$, approximately. It is

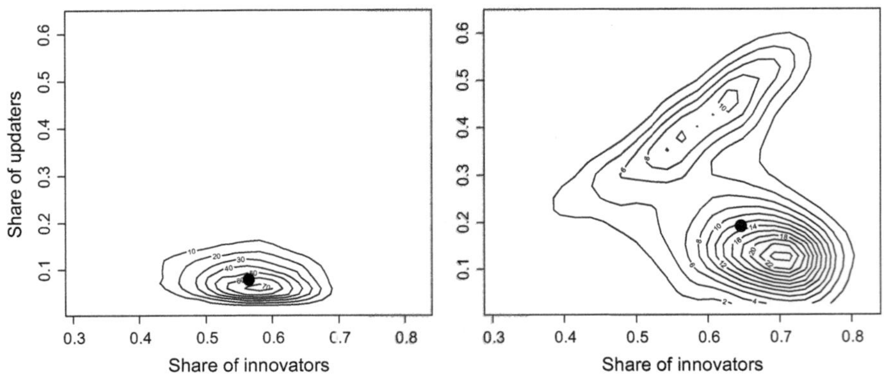

Fig. 4 Contour lines of the simulated joint densities of x^N and y_U^N for $q = 2.8$ (*left panel*) and $q = 3.4$ (*right panel*). *Filled dots* show the median shares, which are also visible in the *right panel* of Fig. 1, relative to $q = 2.8$ and $q = 3.4$

important to stress that, up to some variability discussed previously, all such configurations are equivalent with respect to median revenues.

The pronounced difference between the two panels of Fig. 4 is also indicative of the emergent strategic complexity about the critical value of q. The simple attempt to summarize the results of M simulations is hard in this case, as shown by the somewhat puzzling location of the filled point in the right panel, which is barely representative of the richness and complexity of the outcomes.

Discussion and Conclusion

The findings presented in the previous section allow to deepen the discussion on the rationality of the trade-in campaign of Apple in the US market that was introduced in Sect. 1. Our model singles out that there are values of q in which upgrades should be extremely cheap or even free to boost market shares. In the specific case under scrutiny, at the cost of stretching the model a little bit, we can provide a suggestive description of the following two scenarios: in the first period Apple virtually enjoys a monopolistic position in the market, whereas in the second one the entry of a strong competitor forces Apple to change its pricing campaign. More precisely, we identify period 1 with 2010 and T_1 with iPhone 4, period 2 with 2012/2013 and T_2 with iPhone 5. The competitor we are thinking of is Samsung Electronics Co. with its Android operating system. In April 2013, Bloomberg's Adam Satariano described how Apple's quarterly profit was projected to shrink for the first time in a decade, especially due to this new competitor (see Satariano 2013).

As a first approximation, we model the competition with a simple decrement in the externality parameter q, induced by the entry of the competitor.

(continued)

Let's assume the value of q for Apple in 2010, before the arrival of Samsung, is high (say, close to $q = 4$). This value is justified by the remarkable loyalty Apple's *fans* have always demonstrated to the company. As a consequence, Apple is in a very strong position and can put in place a pricing campaign in which prices are very high, without suffering a decrease in revenues. Once the competitor enters the market (during 2012/2013), a sudden decrease in q may be expected. Suppose indeed, that q falls to about $q = 3.5$. The picture is, thus, different: Apple must revise its aggressive pricing policy in order to maintain its market share. Now, our results suggest that the optimal policy is to give away the update for free. Our model, although very stylized, supports the rationality of an aggressive reduction in the price of updates.

We would like to mention a second feature of the trade-in program implemented by Apple, which is de facto equivalent to a further price reduction for the upgrade: Apple launched the iPhone 5 together with a new release of the iPhone operating system iOS7, which was believed by many users to be the cause of a deterioration in the performances of iPhone 4. It was possible to overcome such "technical" problem by replacing the battery at a cost of about $70, or just upgrading to the new iPhone 5 with a lump-sum of about $99 (thanks to the high discount due to the trade-in program). We do not know whether this was a measure intentionally planned to pursue the company's own objectives (see Rampell 2013) but, in practice, these events additionally increased the cost of keeping the iPhone 4.

On a different note, our model vividly illustrates a feature of emerging equilibria of interacting agents in binary decisions models. Even when a formal analysis rules out the existence of multiplicity, the sampling variability inherent in any simulation can lead to markedly different results. For some values of q, which approximate what would be described as a (near) "tangency" in Granovetter (1978), this is exactly what happens in our setup. The relevance is twofold: on the first hand, outcomes are affected by unavoidable levels of uncertainty, despite the fact of being based on optimal decisions, and the firm should be aware of this strategic unpredictability; on the other hand, such sensitivity does not disappear even if an analytical model is at hand. For instance, existence and uniqueness of an equilibrium do not imply that finite size simulations will converge to the unique equilibrium.

We assume in the model that the individual shocks related to the preferences of the agents are independent. This is a limitation of the present treatment as it is likely that ε_1 is orthogonal to ε_2 only if the two competing technologies are radically different. In many realistic cases, subsequent waves of products may carry relatively minor technical changes or moderate improvements in usability. In these cases, the noise terms would be (strongly) positively correlated. Hence, our model may be more suited to describe

(continued)

situations in which fundamental developments have been introduced or in the presence of a notable shift with respect to the past paradigms.

Further generalizations of the model may also remove the assumption that the firm commits itself to fix prices in the first period. Clearly, the price of getting (or updating to) T_2 needs not to be disclosed to customers in the first period but an alternative course of action would suggest to determine p_{2A} and p_{2U} *after* the market share x^N emerges. Instead of being unconditionally worked out together with p_1, the selection of prices for T_2 should be conditional on the realized market share, in a backward-like fashion resembling Bellman dynamic programming principle. In this respect, the revenues of the present model can be interpreted as lower bounds for a conditional pricing strategy, which may result in additional marginal excess gains.

Acknowledgements The authors thank Marco LiCalzi for insightful comments and discussions and participants to seminars held at XXXVII AMASES meeting, Stresa; Collegio Carlo Alberto, Torino and Alpen-Adria Universität, Klagenfurt. We received financial support from MIUR under grant "Robust decision making in markets and organizations" (PRIN20103S5RN3).

References

Adner R, Levinthal D (2001) Demand heterogeneity and technology evolution: implications for product and process innovation. Manag Sci 47(5):611–628

Brock WA, Durlauf S (2001) Discrete choice with social interactions. Rev Econ Stud 68(2):235–260

Dai Pra P, Sartori E, Tolotti M (2013) Strategic interaction in trend-driven dynamics. J Stat Phys 152:724–741

Dawid H (2006) Agent-based models of innovation and technological change. In: Tesfatsion L, Judd KL (eds) Handbook of computational economics, vol 2. Elsevier, Amsterdam, pp 1235–1272

Deffuant G, Huet S, Amblard F (2005) An individual-based model of innovation diffusion mixing social value and individual benefit. Am J Sociol 110(4):1041–1069

Granovetter M (1978) Threshold models of collective behavior. Am J Sociol 83(6):1420–1443

Kiesling E, Gunther M, Stummer C, Wakolbinger LM (2012) Agent-based simulation of innovation diffusion: a review. Cent Eur J Oper Res 20(2):183–230

Nadal J, Phan D, Gordon M, Vannimenus J (2005) Multiple equilibria in a monopoly market with heterogeneous agents and externalities. Quant Financ 5(6):557–568

Rampell C (October 29, 2013) Cracking the apple trap. The New York Times Magazine

Satariano A (April 22, 2013) Apple profit probably fell amid growth slowdown for iphone. Bloomberg.com

Evaluating Scenarios for Upgrading Sustainability of the Meat Supply Chain

Eva van den Broek and Tim Verwaart

1 Introduction

Consumer demand for organic meat has been increasing in Europe and the US by 300 % between 1999 and 2007 (Sahota 2009); in 2012, the increase in organic meat in the Netherlands was 48.2 %; nevertheless, the share of organic meat remains at a very low level around 3 % (LEI Monitor Duurzaam Voedsel 2013). Farmers are struggling with the low coverage for regular meat production. Since their profits have been below zero in 80 % of the months since January 2006 in the Netherlands (LEI Bedrijven InformatieNet 2013), producers are eager to embrace sustainable farming as a way to switch from cost-driven to value-driven products.

Despite the alarming state of the primary sector, new business models involving sustainable production have so far not managed to capture a large market share. Reasons for this mismatch are diverse. First, the meat supply chain is characterized by short term markets, while investing in sustainability certification only pays off after a time interval. Moreover, hardly any brands are developed at the producer level, pushing the competition towards price competition only (de Jonge and van Trijp 2013). Finally, the inherently dynamic dependencies between consumer buying behaviour and the availability of sustainable meat in the supermarket pose an additional barrier on the uptake of sustainable meat production.

Previous models on artificial markets have added to the understanding of realistic features such as local interaction, learning and time dynamics (see f.i. Kirman (2008); for an overview of supply chain ABMs, see Mizgier et al. 2012). Agent-based simulations allow for differentiation of the actors' characteristics and the diffusion of social norms. This may considerably affect the overall sustainability

E. van den Broek (✉) • T. Verwaart
LEI Wageningen UR, Postbus 29703, 2502LS Den Haag, The Netherlands
e-mail: e.vandenbroek@wur.nl; tim.verwaart@wur.nl

F. Amblard et al. (eds.), *Advances in Artificial Economics*, Lecture Notes in Economics and Mathematical Systems 676, DOI 10.1007/978-3-319-09578-3_19

levels. Fluctuations induced by consumer demand or batch deliveries have been shown to induce bullwhip effects and to propagate through the chain up to producer bankruptcies (see f.i. Lacagnina and Provenzano 2010). To increase the external validity of such models, types of (strategies of) traders and producers have been derived based on interviews, surveys or other real life data (Rouchier 2004). Indeed, some models even allow for interaction between software and human agents (Meijer et al. 2011).

Our setup and scenarios are informed by a research project conducted in 2013, which comprised workshops and interviews with experts, stakeholders and researchers of the Dutch pork and poultry production and retail chain (Reinders et al. 2014). Based on successful transitions towards sustainability in other sectors (among others horticulture, the veal industry, and the Dutch coffee and soy markets) a number of plausible business models were developed and presented to the stakeholder parties. In two consecutive rounds, future scenarios were constructed that may lead to an increase in value and sustainability, but require coordination. These scenarios can be characterized as either modular or captive supply chains (Gereffi et al. 2005).

The scenarios were developed along two trends: increasing brand differentiation and transparency towards consumers vs extensive cost reduction through chain internalization of external costs. Although the policy recommendations are specific to a small national market and focus on the challenges faced by Dutch primary producers, the scenarios are applicable to a broader set of food production chains in which the transition towards sustainability is required by government and NGOs, but hampered by fierce price competition between brands.

The aim of this paper is to construct a model that incorporates the above dimensions and to observe the positioning of consumers, producers and brands over time in a series of business scenarios. We apply agent-based simulation because we want to investigate the interactions and diversity among actors and their effects on the transition to sustainability. Specifically, we want to address the following research questions: How do the scenarios differ with respect to the speed of uptake of sustainable meat consumption and producer welfare? Which plausible scenario is best from a perspective that takes into account both overall sustainability levels and producer welfare? We hypothesize that the interaction between consumer demand, shifting norms and market dynamics leads to large differences in producer defaults.

2 The Model

In this section the agents, their interactions and their typology are described. The agents act in an environment where a steady supply of regular, conventionally produced meat is ensured. In the beginning of the simulation, only regular meat is supplied and NGOs start campaigning for sustainability among consumers and producers. The supply chain is represented by a set of brands. Each brand offers meat according to a certified level of sustainability, which may range from

100 % regular to 100 % organic. The model assumes that consumers are willing to pay a premium for sustainable meat, in response to the NGOs' campaigns and consequently evolving social norms. For supply of sustainably produced meat, the brands can pay the producers a premium. Producers may decide to invest in their production system in order to switch from regular to certified sustainable production. However, if the supply of sustainable meat exceeds the demand, the sustainable producers must sell the surplus for the price of regular meat.

The agent-based model aims to simulate the dynamics of this system for a period of several years, with time steps of 1 week, under several regimes of market organization and information supply. Observable outputs relate to the consumption of sustainably produced meat, the level of sustainability of meat production, the distribution of wealth among farmers, the number of farms defaulting due to overproduction, and the distribution of the turnover of brands.

Agents The model contains four populations of actors: consumers, producers, brands, and an information agent representing a nongovernmental organisation promoting sustainability and animal welfare (see Fig. 1). Each actor is characterized by its individual preference on a continuous sustainability scale.

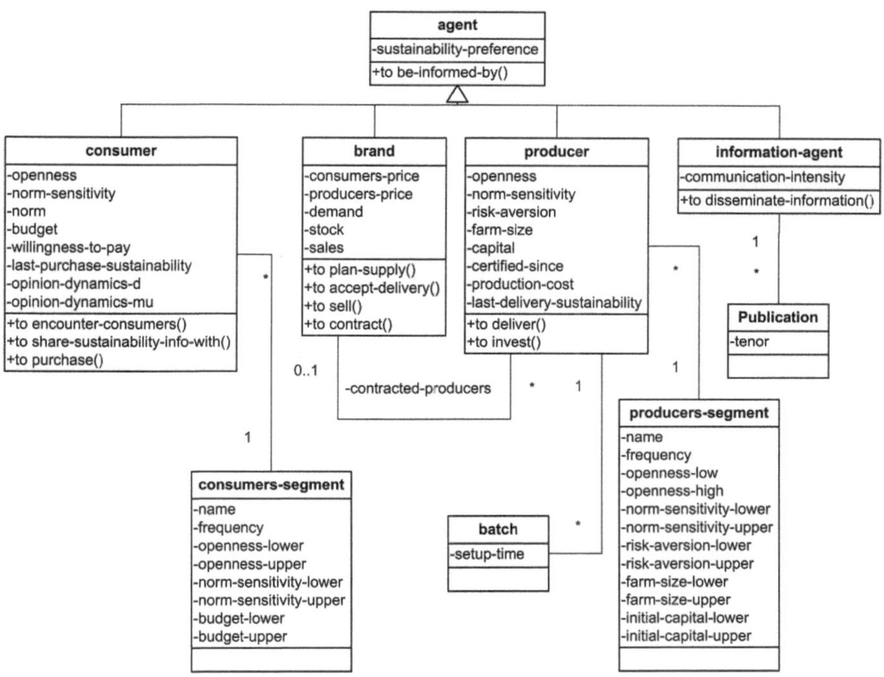

Fig. 1 Class diagram representing the agent types, data structures, and methods in the simulation

Consumers The cognitive architecture of consumers allows for non-rational behaviour. Their sustainability preference is the result of their openness to communications and the sustainability preference and tenor of received messages from information agents they are informed by. They maintain a belief about the social norm for sustainable behaviour through dynamic opinion formation (Deffuant et al. 2000), sharing information about their last purchase with other consumers they encounter. This noisy belief about the "norm" and their own sustainability preference determine the premium the consumer is willing to pay for an organic product, within the consumer's budget limit. Formally, the Willingness-to-Pay of a consumer i at time t is computed as

$$\mathrm{WTP}_{i,t} = \min\{\mathrm{budget}_i, (1 - \mathrm{NormSens}_i) * \mathrm{STP}_{i,t} + \mathrm{NormSens}_i * \mathrm{Norm}_{i,t}\} \quad (1)$$

where budget_i is the budget available to consumer i, $\mathrm{NormSens}_i$ is the norm sensitivity of consumer i, $\mathrm{STP}_{i,t}$ is the sustainability preference of consumer i and $\mathrm{Norm}_{i,t}$ is the consumer's belief about the norm at time t. Given that a brand exists with a consumer price that lies within the consumers budget, consumer i buys a product from the brand B with sustainability preference STP_B, satisfying

$$B_{i,t} = \arg\min_b |\mathrm{WTP}_{i,t} - \mathrm{STP}_b| \wedge \mathrm{STP}_b \leq \mathrm{budget}_i \wedge \mathrm{stock}_b > 0. \quad (2)$$

Brands Brands rationally optimize their turnover given the constraints of demand and supply and are positioned on the sustainability spectrum. Their position on the latter spectrum is indicated by their sustainability preference. The brands operate with a fixed consumer premium, which is proportional to their sustainability preference. They source a mix of regular and sustainable meat (for which they pay a premium), again proportional to their sustainability preference. Every week they try and source an integer number of batches of sustainably produced pigs in order to bring their stock at least at the level of the previous week's sales, plus one batch of pigs. The brands have the capability to close long term contracts with producers for sourcing sustainably produced bigs, but they apply this capability only in particular business scenarios.

Producers Producers are influenced only by the information agent and brand demand. They deliver goods that are either certified or not. Producers set up batches of pigs for fattening. The size of their farm expresses the number of batches they can house. If there is room, they set up new batches, which are delivered after 20 weeks. Producers receive an initial capital. They may decide to invest in a sustainability certificate. If they do so, their production cost will be increased for every batch they set up. If their capital allows for investment, a producer's decision to invest in a certificate depends on their risk aversion, on the price premium at time t, their sustainability preference, and on the demand for sustainable pork expressed by the brands. They decide as follows:

$$\text{certifiy}_{j,t} = \begin{cases} 0 & \text{if } t = 0 \\ 1 & \text{if } t > 0 \wedge (p_t - c)/p_t * \text{STP}_j * D_t/100 > \text{RA}_j^3 \\ \text{certifiy}_{j,t-1} & \text{otherwise} \end{cases} \quad (3)$$

where p_t is the premium for certified produce at time t, c the production cost, STP_j the sustainability preference of producer j, D the unsatisfied demand at time t and RA_j the risk aversion of producer j. Only 20 weeks after the investment the first sustainable batches can be delivered and the producer may receive the sustainability premium. The farmers sell their sustainably produced batches to the brand with a demand that offers the highest producer premium, unless they have a long term contract with a brand. If they cannot sell their sustainable produce because of insufficient demand, they must dump it on the regular market and lose money. If they run out of capital, they revert irrevocably to regular production.

Information Agent The information agent influences the consumers and producers by sending messages with a particular intensity and tenor (the information agent's sustainability preference). In each time step consumers and producers receive the messages with a probability that equals the information agent's communication intensity, upon which they update their sustainability preference according to the following formula:

$$\text{STP}_{i,t} = \begin{cases} 0 & \text{if } t = 0 \\ (1 - \text{openness}_i) * \text{STP}_{i,t-1} + \text{openness}_i * \text{tenor}_t & \text{if } t > 0 \end{cases} \quad (4)$$

where $\text{STP}_{i,t}$ stands for the ith consumer's or producer's sustainability preference at time t and openness_i for its susceptibility for information.

Typology Both consumers and producers are characterized as types. Consumer types are classified following (Hessing-Couvret and Reuling 2002) as conservative, caring, balanced, engaged or openminded; producers according to de Lauwere et al. (2002) as traditional, economical, balanced, broad-minded or professional (see Table 1). Apart from their sustainability preference, consumer types differ on the dimensions openness, sensitivity to social norms, and in their budget constraint. These parameters, the information and the social norm they experience in their network, together determine their Willingness-to-Pay, or the premium they are willing to pay for organic products in a specific time step. Producer types vary in their openness, risk attitude and capital, which together determine their binary decision to invest in certificates or not. In the present simulation we assume equal farm size so that producers deliver one batch per week.

Table 1 Typology of producers and consumers

Consumer	Conservative	Caring	Balanced	Engaged	Openminded
Frequency	0.27	0.15	0.21	0.18	0.19
Openness	Low	Low	Med	High	High
Budget	Low	Low	Med	High	High
Norm sensitivity	High	Low	Med	High	Low
Producer	Traditional	Economical	Balanced	Professional	Openminded
Frequency	0.22	0.14	0.21	0.25	0.18
Openness	Low	Low	Med	High	High
Norm-sensitivity	High	High	Med	High	Low
Risk aversion	High	High	Med	High	Low
Farm size	20	20	20	20	20
Capital	50,000	50,000	50,000	50,000	50,000

Values were randomly generated within a certain range. "High" denotes a value between [0.65–0.95]; "med" between [0.35–0.65]; and "low" between [0.05–0.35]. Similarly, a high budget is set to 1, a "med" budget to a value between [0–1], and low to [0–0.5]

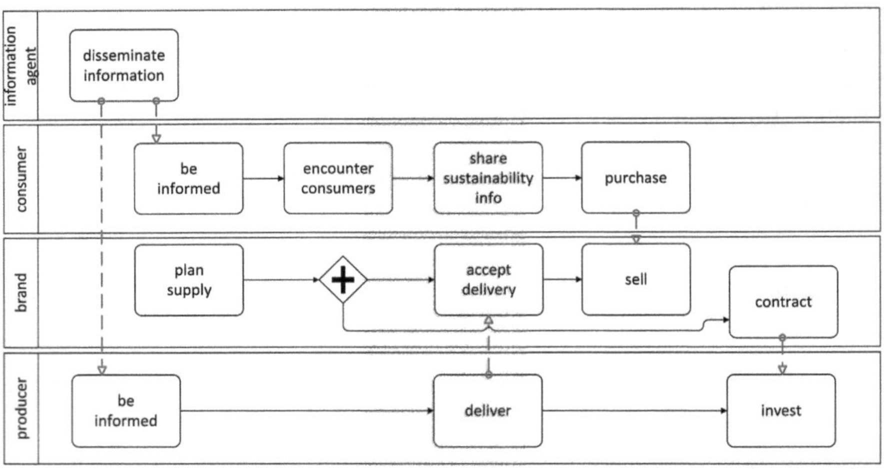

Fig. 2 Activities performed by the agents in each time step

Time Steps Figure 2 presents an overview of the agent's activities in each time step. At the beginning of a time step (roughly equivalent to a week) the information agent provides information that may or may not nudge the consumers and producers towards a stronger or weaker sustainability preference. Brands decide how much to source, based on consumer demand and meat supply in the previous time step. Producers deliver their mature batches. If demand for certified produce is low, they may be forced to sell for regular prices on the world market. They decide whether or not to invest in a certificate if their capital allows for investment; in some business scenarios a long term contract must be closed with a brand. Consumers encounter other consumers, exchange information about their last purchase and then purchase based on their Willingness-to-Pay and the consumer price.

Controls and Initialisation Simulation time is set to 312 weeks, during which each consumer encounters another 20 times. In the present simulation we assume equal farm size, allowing for the delivery of one batch of pigs per week by each producer. Production costs are set to 1,000; the production premium for sustainable meat is set to 2,000. The premium consumers pay for a completely sustainable product is set to 3. Opinion dynamics are set to 0 and the norm is initialised at 0. For other parameters of the consumers and producers, see Table 1. There is one information agent with a sustainability preference of 1 and a communication intensity of 0.05, which means that in every time step one out of 20 agents is influenced at all by the information.

Scenario Description We compare four scenarios against a baseline scenario, based on the Dutch pig and poultry production systems (Reinders et al. 2014). The scenarios, inspired by developments towards sustainability in other sectors, represent a subset of the settings that can be implemented in the model. In the baseline scenario, no coordination exists and the market contains only regular and organic meat products.

In Scenario A (market differentiation), we allow for 10 brands with intermediate levels of sustainability, varying between a 10 % and 90 % share of organic meat. This scenario reflects the current situation in the Netherlands, with a set of intermediate brands competing for market share.

In Scenario B (Green Track), inspired by the soy trade, one intermediate brand is introduced in the baseline scenario, containing the optimum proportion of organic meat (10–99 %) given the supply and the consumers WTP. This reflects a situation in which the supply chain guarantees to consumers that a minimum percentage of the meat is certified. This percentage is raised as the WTP increases. It is a stylized version of a cooperative approach that tackles the optimal carcass valorisation, one of the major obstacles to sustainability in the chicken production chain.

Scenario C builds on A. Here, a commercial market orientation platform confers supply forecasts for organic meat to the producers, who take the information into account in their investment decisions. In horticulture, such a market platform has shown to function as a catalyst for the rise of certification standards, by negotiating with NGOs, sharing good practices and finetuning the supply to demand.

Scenario D (producers' organisation) builds on B, allowing for contracts between a brand and a group of producers with a fixed premium for a certain amount of certified meat. This reflects a shift from short term to long term contracts in order to lift the risk of demand and supply uncertainty off the producers. Since a producers' organisation may evolve in reality into a broader institution, such as a bargaining association or a first port of call for the retail, it may lead to further chain integration, as is the case in the Dutch veal sector and the German and French poultry sector.

3 Results

We ran simulations in Netlogo (Wilenski 1999) for a period of 312 weeks (6 years) with equal parameter settings for consumers and producers in all scenarios.[1]

Figure 3 summarizes outcomes of an average run for each of the five scenarios. The top row shows the development of the brands' demand for certified sustainable meat as a percentage of the total demand. The demand can remain unsatisfied for some time due to the time lag in the production. The second row of graphs shows the share of the brands' demand that cannot be satisfied at the current production level. Unsatisfied demand challenges producers to invest in sustainable meat production. The total number of currently active producers of certified sustainable meat is reported in the third row of graphs. Sustainable production entails an increased cost level, which reduces the producers' capital. A profit is made when the sustainable produce can be sold as such, but the cost is not recovered if the sustainable

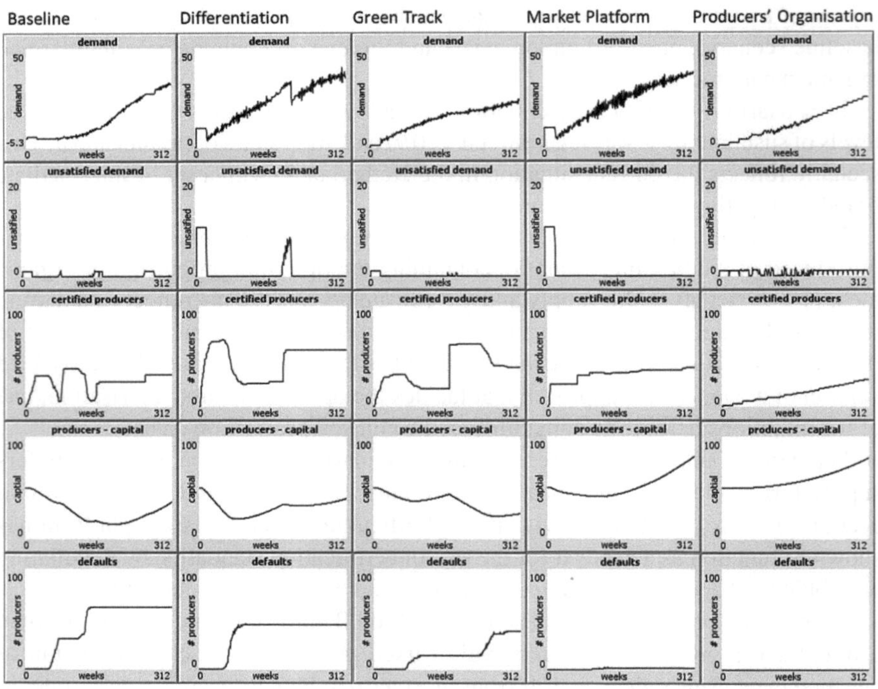

Fig. 3 Overview of results of the baseline simulation and four alternative scenarios; for each scenario the graphs present the evolution of total demand for sustainable meat from the brands, the demand that can not be satisfied at the current production level, the number of certified producers, the total capital of all producers, and the accumulated number of defaults

[1]The NetLogo simulation is available for download from http://www.verwaart.nl/Sustainability.

produce must be sold as surplus in the regular market. The fourth row displays the development of the total capital of all producers. Some farmers may lose their investment in sustainable production, in which case they must revert to regular production. The accumulated number of such defaults is reported in the bottom row of graphs in Fig. 3. The following paragraphs discuss the outcomes of each scenario.

In the *baseline scenario*, consumers are offered the choice to purchase either regular or 100 % organic meat. For organic meat they must be willing to pay a price premium. When the first demand develops under pressure of NGOs, too many producers are challenged to invest in organic meat production with the prospect of the price premium. Due to the pork cycle effect, the supply of organic meat largely exceeds the demand when the first batches are delivered. The surplus must be sold on the market for regular meat, without price premium. Many producers who invested in sustainable production lose their investment and must revert to regular production. This pattern is repeated when demand further develops under pressure of NGOs and shortage of sustainably produced meat occurs. Few producers survive and then make a good profit under increasing demand. A skewed distribution of capital among the producers results. Figure 4 displays the distribution of capital per scenario. In the baseline scenario, only consumers from the "Engaged" and "Openminded" segments purchase sustainable meat (see Table 2), because the premium for 100 % organic meat is beyond the Willingness-to-Pay of the other consumers.

Compared with the baseline scenario, the development of demand is considerably accelerated in the *differentiation scenario (A)*. Immediately when a slightly increased Willingness-to-Pay results from NGO campaigns, early adopters among the consumers can purchase meat from brands with a slightly increased level of sustainability. When the Willingness-to-Pay further increases, these consumers can buy meat with higher levels of sustainability. Meat with various levels of

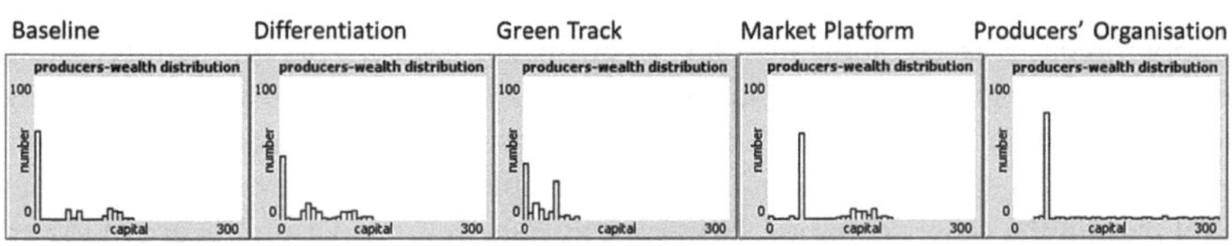

Fig. 4 Distribution of capital among producers, as it emerges after six simulated years

Table 2 Sustainable meat purchase by consumer segment after 6 years per simulated scenario

Scenario	Conservative	Caring	Balanced	Engaged	Openminded
Baseline	No	No	No	Yes	Yes
Differentiation	Yes	Yes	Yes	Yes	Yes
Green track	No	No	No	Yes	Yes
Market platform	Yes	Yes	Yes	Yes	Yes
Producers' org.	No	No	No	Yes	Yes

sustainability and according price levels is available. In contrast with the baseline scenario, consumers from all segments can afford sustainable meat, to a level that matches their budgets.

Like in the baseline scenario, a pork cycle effect occurs in the beginning of the differentiation scenario simulation. Many early adopting producers lose their investment. Those who survive must initially dump some portion of their produce on the regular market, but revenues increase as the demand evolves. Return is shared only among the organic producers, so in the end the surviving organic producers make a good profit and a skew distribution of capital emerges (see Fig. 4). After several years serious shortage occurs, and more farmers invest in sustainable production. Because of the well-developed and further evolving demand, this does not lead to additional defaults.

As in scenario *A*, the simulation outcomes for the *Green Track scenario (B)* show a more rapid evolution of demand for sustainable meat, but after 6 years the total sustainability reaches a similar level as in the baseline scenario. Furthermore, as in the baseline scenario, sustainable consumption is eventually limited to particular consumers segments (see Table 2). Only "Engaged" and "Openminded" consumers are willing to pay the high premium for 100 % organic meat, whereas scenario *A* attracts consumers from all segments. As a consequence, a lower level of total sustainability (i.e. number of certified producers) than in scenario *A* is attained in the Green Track scenario.

Since fewer consumers buy sustainable meat, the total revenue is lower than in scenario *A*. The number of defaults is high, as in the previous scenarios. From the producers' viewpoint, scenario *B* is less attractive than scenario *A*, but still more attractive than the baseline: more producers survive the initial pork cycles. The resulting distribution of capital is similarly skewed as in scenario *A*, but has a lower average.

The *market platform scenario (C)* adds a market orientation platform to scenario *A*, with the purpose to buff the pork cycle effects. All producers have access to supply forecasts based on the number of certificates issued. This supply forecasting results in more gradual development, which has the potential of greatly reducing the number of defaults. As in scenario *A*, all consumer segments adopt sustainable consumption to some degree, and total sustainable production evolves to a higher level than in the other scenarios. However, average returns per producer are lower, because the additional revenues from sustainable farming are shared among more producers.

The supply planning with the brands as intermediaries entails a rather strong bullwhip effect (as described in Lacagnina and Provenzano 2010) in scenario *C*. The shortage arising after some years in the simulation causes fluctuations in consumer demand, which are reinforced in the supply chain. When implementing scenario *C*, measures could be desired to reduce the bullwhip effect (see, f.i., Moyaux and McBurney 2006).

In the *producers' organisation scenario (D)* a group of producers closes exclusive contracts with a brand. While in scenario *B* the brand only attunes the level of sustainability to the market opportunities, in scenario *D* it also manages the

Table 3 Characteristics of the simulation outcomes per scenario

Scenario	Sustainability (rank)	Consumer uptake	Producer defaults	Average revenue	Capital distribution
Baseline	3–4	Elite	Highest	Low	Very skew
Differentiation	2	Broad	High	Moderate	Skew
Green track	3–4	Elite	High	Moderate	Skew
Market platform	1	Broad	Low	High	Moderate
Producers' org.	5	Elite	None	Highest	Very skew

production of sustainable meat. The contracts eliminate the producers' risks. The brand contracts new farmers only when consumer demand has increased sufficiently. Due to the time lag in production, a slight shortage persists. As a result, the share of sustainable production proceeds slower than in the other scenarios.

Scenario *D* entails high revenues for producers who are incorporated into the association and prevents defaults due to the pork cycle effect among the regular producers. As a result, both the average capital and the skewness of the distribution of capital are high. As in the baseline and scenario *B*, the brand focuses on the consumer segments with high Willingness-to-Pay. Combined with the managed introduction of sustainable meat, this results in a low level of total sustainability in the supply compared to other scenarios.

The simulation outcomes are summarized for comparison of characteristics across scenarios in Table 3. The results suggest that the scenarios greatly differ with respect to the progress of sustainable meat consumption and its effects on producer welfare. The uptake of sustainable produce is bound by the Willingness-to-Pay of the various consumer types. In the long run, the differentiation scenario offers the highest sustainability levels, since it caters to all consumer segments. Total producer capital plunges after initial investment in the first three scenarios, due to classic pork cycle effects. In the market platform and the producers' organisation scenario not nearly as many defaults occur, but the distribution of capital is skewed; a subset of producers reaps the benefits from the sustainable meat production.

Conclusion

Sustainability in the meat supply chain depends to a large extent on coordination between the supply chain actors. This paper presents a set of multi-agent simulations of the meat market in which the interactions between producers, brands, NGOs and consumers are modeled. Four plausible scenarios based on developments in other fresh produce sectors are implemented and compared with respect to the speed of uptake of sustainable meat production and consumption.

(continued)

Our model suggests that the interaction between consumer demand, shifting norms and market dynamics leads to large differences in welfare between the four scenarios. From a perspective that takes into account both overall sustainability levels and producer welfare, the market platform scenario appears most desirable. In our stylized setup, the consumer surplus is evenly divided between brands and producers. In reality, interactions between supply chain partners from retail and meat industry will influence the market power and the division of surplus. Despite this simplification the model offers valuable insights for the chain actors. For instance, from a producer's perspective initiating or joining a producers' organisation brings the benefits of being a first adopter, in addition to inducing a shift in the market towards a less risky and more profitable production environment. It may however result in a slower supply of sustainable produce to consumers. By finetuning their level and tenor of information dissemination, an NGO may be able to inflate or dampen the cycle effects in all scenarios. Surprisingly, their efforts may be most effective in the market oriented scenarios.

The model allows for evaluation of strategies that aim to balance competition and coordination in the development of sustainable supply chains. It captures the interplay between consumer demand, market dynamics, societal pressure, and perspectives for producers and supply chain partners. By doing so, the simulation offers insights to policy makers and supply chain actors in designing long term strategies towards sustainability.

References

de Jonge J, van Trijp H (2013) Meeting heterogeneity in consumer demand for animal welfare: a reflection on existing knowledge and implications for the meat sector. J Agric Environ Ethics 26:629–661

de Lauwere C, Verhaar K, Drost H (2002) Het mysterie van het ondernemerschap. Rapport 2002-2. IMAG, Wageningen (in Dutch)

Deffuant G, Neau D, Amblard F, Weisbuch G (2000) Mixing beliefs among interacting agents. Adv Complex Syst 3:87–98

Gereffi G, Humphrey J, Sturgeon T (2005) The governance of global supply chains. Rev Int Polit Econ 12:78–104

Hessing-Couvret E, Reuling A (2002) Het WIN-model Waardensegmenten in Nederland. NIPO, Amsterdam (in Dutch)

Kirman A (2008) Artificial markets: rationality and organisation. In: Schredelseker K, Hauser F (eds) Complexity and artificial markets. Lecture notes in economics and mathematical systems, vol 614. Springer, Heidelberg

Lacagnina V, Provenzano D (2010) Threshold rule and scaling behavior in a multi-agent supply chain. In: LiCalzi M et al (eds) Progress in artificial economics. Lecture notes in economics and mathematical systems, vol 645. Springer, Heidelberg

LEI Bedrijven InformatieNet (2013) www.wageningenur.nl/lei (in Dutch)

LEI Monitor Duurzaam Voedsel (2013) www.wageningenur.nl/lei (in Dutch)

Meijer SA, Ragothoma J, King R, Palavalli B (2011) Indian food supply chains: a game and model to study economic behavior. In: Osinga S et al. (eds.) Emergent results of artificial economics. Lecture notes in economics and mathematical systems, vol 652. Springer, Heidelberg

Mizgier KJ, Wagner SM, Holyst JA (2012) Modeling defaults of companies in multi-stage supply chain networks. Int J Prod Econ 135:14–23

Moyaux TP, McBurney P (2006) Reduction of the bullwhip effect in supply chains through speculation. In: Bruun C (ed) Advances in artificial economics. Lecture notes in economics and mathematical systems, vol 584. Springer, Heidelberg

Reinders M, Poppe K, Immink V, van den Broek E, van Horne P, Hoste R (2014) Waardevolle perspectieven voor vlees. LEI Wageningen UR, Den Haag (in Dutch)

Rouchier J (2004) Interaction routines and selfish behaviours in an artificial market. In: WEHIA, Workshop of economics with heterogenous interacting agents, Kyoto, 29–31 May 2004

Sahota, A (2009) The global market for organic food & drink. In: The world of organic agriculture. Frick and Geneva, Bonn. ITC/FIBL/IFOAM

Wilenski, U (1999) NetLogo. http://ccl.northwestern.edu/netlogo/. Center for Connected Learning and Computer-Based Modeling, Northwestern University, Evanston

MIX
Papier aus verantwortungsvollen Quellen
Paper from responsible sources
FSC® C105338

If you have any concerns about our products,
you can contact us on
ProductSafety@springernature.com

In case Publisher is established outside the EU,
the EU authorized representative is:
Springer Nature Customer Service Center GmbH
Europaplatz 3, 69115 Heidelberg, Germany

Printed by Libri Plureos GmbH
in Hamburg, Germany